Integration of Science and Technology with Development
(Pergamon Policy Studies-22)

Pergamon Titles of Related Interest

Carman   *Obstacles to Mineral Development: A Pragmatic View*

Diwan/Livingston   *Alternative Development Strategies and Appropriate Technology: Science Policy for an Equitable World Order*

Francisco/Laird/Laird   *The Political Economy of Collectivized Agriculture: A Comparative Study of Communist and Non-Communist Systems*

Golany   *Arid Zone Settlement Planning: The Israeli Experience*

Goodman/Love   *Management of Development Projects: An International Case Study Approach*

Laszlo/Baker/Eisenberg/Raman   *The Objectives of the New International Economic Order*

Meagher   *An International Redistribution of Wealth and Power*

Morris   *Measuring the Condition of the World's Poor: The Physical Quality of Life Index*

The Rothko Chapel   *Toward A New Strategy for Development*

Stepanek   *Bangladesh—Equitable Growth?*

PERGAMON
POLICY
STUDIES

Integration of Science and Technology with Development/Caribbean and Latin American Problems in the Context of the United Nations Conference on Science and Technology for Development

Edited by
D. Babatunde Thomas
Miguel S. Wionczek

Published in cooperation with
Florida International University,
The Institute of Social and Economic Research,
University of the West Indies and the
Institute of Development Studies,
University of Guyana

Pergamon Press
NEW YORK • OXFORD • TORONTO • SYDNEY • FRANKFURT • PARIS

Pergamon Press Offices:

U.S.A. Pergamon Press Inc., Maxwell House, Fairview Park, Elmsford, New York 10523, U.S.A.

U.K. Pergamon Press Ltd., Headington Hill Hall, Oxford OX3 0BW, England

CANADA Pergamon of Canada, Ltd., 150 Consumers Road, Willowdale, Ontario M2J, 1P9, Canada

AUSTRALIA Pergamon Press (Aust) Pty. Ltd., P O Box 544, Potts Point, NSW 2011, Australia

FRANCE Pergamon Press SARL, 24 rue des Ecoles, 75240 Paris, Cedex 05, France

FEDERAL REPUBLIC OF GERMANY Pergamon Press GmbH, 6242 Kronberg/Taunus, Pferdstrasse 1, Federal Republic of Germany

Library of Congress Cataloging in Publication Data
Main entry under title:

Integration of science and technology with development.

(Pergamon policy studies)
Proceedings of a symposium held in Miami, Apr. 6-8, 1978, sponsored by Florida International University, the Institute of Social and Economic Research, University of the West Indies, and the Institute of Development Studies, University of Guyana.
Includes index.
1. Underdeveloped areas—Technology—Congresses.
2. Technology—Latin America—Congresses. 3. Technology transfer—Congresses. I. Thomas, D. Babatunde.
II. Wionczek, Miguel S. III. Florida International University. IV. Mona, Jamaica. University of the West Indies. Institute of Social and Economic Research.
V. University of Guyana. Institute of Development Studies.
T49.5.I55 1979   338.91'8   78-26984
ISBN 0-08-023881-5

Printed in the United States of America

Contents

Introduction

A future historian will note with all probability that the second half of the twentieth century was characterized by two seemingly contradictory trends: the accelerated scientific and technological progress in the industrialized part of the world economy, and the generalized socio-economic stagnation in the rest of the globe.

The persistent coexistence of these two phenomena gave rise in the 1960s to a worldwide debate on the subject of mobilizing science and technology for the solution of the problem of underdevelopment. While, as one of the papers included in this volume points out, the debate translated itself into the staggering output of books, studies, and policy-oriented proposals – at national and international levels – not only did little actual application of science and technology for development occur, but all sorts of gaps between the rich and the poor countries continued to increase. This disappointing situation resulted in questioning the very relevance of modern science and technology for the poor societies, and in the emergence of an idea now widely known as "appropriate technology."

In the midst of these growing intellectual conflicts another idea started gaining ground in the world scientific community in the early 1970s. This idea is that the divorce between science and technology on the one hand, and development problems on the other, warranted an international conference that would hopefully bring to an end this unsatisfactory state of affairs. The political support offered for this idea by the governments of both the poor and the rich countries resulted in the forthcoming United Nations Conference on Science and Technology for Development(UNCSTD) scheduled for August 1979 in Vienna, Austria.

Those who had a chance to follow the initial preparations for UNCSTD in 1976-77, and were able to analyze the results of the first meeting of its Preparatory Committee, held in New York in February 1977, discovered that these preparations were highly politicized and devoid of substance, and were transformed immediately into a dialogue between scientists from the developed countries and the coalition of diplomats and bureaucrats from the less developed countries (LDCs).

Since the records of the UNCSTD preparatory meetings strongly suggest that neither group has been particularly conversant with science and technology problems in the context of underdevelopment, there is a high probability that UNCSTD itself may result in another harmful and unnecessary confrontation between the rich and the poor countries. Consequently, in many parts of the world voices were heard that such political confrontation should be avoided, if only because, for a long time, the LDCs will need large amounts of the scientific knowledge and technical know-how already accumulated in the developed countries. This position, supported by practically all scientific and technological communities around the globe, seems to be shared also by many people concerned with science and technology policy issues, in both the rich and the poor countries.

As the preparations for UNCSTD continued to advance in 1977, it became clear that a frustrating international confrontation in Vienna might be avoided if, among other things, the science and technology (S&T) needs of the LDCs were defined, and some mutually acceptable rules for international S&T cooperation established. The definition of the LDCs' needs was necessary because of the magnitude and scope of scientific and technological advancements achieved in the past few decades, but concentrated in the rich countries. An agreement on basic principles of international S&T cooperation was highly advisable because the traditional forms of cooperation have not worked. Many nongovernmental organizations concerned with scientific and technological matters decide to join the debate on these key issues for the purpose of enhancing the probability of UNCSTD's success.

In the fall of 1977, a group of people working on the problems of science and technology policy in the Caribbean and Latin America came to the conclusion that it might be useful — in addition to intergovernmental UNCSTD preparatory meetings on national, subregional, and regional levels — to organize a small academic symposium that would address itself to the whole agenda of the United Nations Conference. Preliminary meetings on the planning of the symposium were held in Mexico City, Miami, and Mona, Jamaica. Considering that such an event fell within the sphere of its own academic concern, Florida International University, a member of the State University System of Florida, offered to sponsor the symposium jointly with the Institute of Social and Economic Research, University of the West Indies, and the Institute of Development Studies, University of Guyana. The full proceedings of the Symposium, held at the FIU campus in Miami in early April 1978, and enlarged by a few additional contributions, appears in this volume.

The Symposium was, in fact, one of the first international academic meetings in which a group of people had an opportunity to discuss science and technology problems in the developing countries of the Western hemisphere as individuals, and not as government officials or international civil servants. The participation of representatives of various United States agencies and academic research centers involved directly or indirectly in UNCSTD, or interested in its outcome, broadened the scope of the debate. Hence, the value of the Symposium for the United States, whose scientific and technological policies have considerable impact upon Latin America and the Caribbean, was perhaps increased.

The agenda focused on four basic propositions that have recently gained acceptance among students of the problems of science and technology for development and among policymakers in the same field:

1. the building up of a minimum local capability to produce scientific knowledge and technological know-how is the precondition for the successful application of science and technology to the solution of difficult and intricate problems of underdevelopment;

2. the task of establishing a local science and technology capability in the developing countries must start with strengthening S&T infrastructure, and improving or redesigning the mechanisms of technology transfer from the advanced countries;

3. there is a need to know more about concrete technological experiences and difficulties of individual LDCs and, in particular, of small LDCs, and

4. the time was ripe to take stock of the achievements and failures of numerous science and technology policy agencies established in Latin America and the Caribbean in the 1960s and the early 1970s.

Finally, it was thought that the symposium offered an opportunity to look with frankness, objectivity, and expertise into the prospects of UNCSTD.

The five parts of the volume deal respectively with the issues arising from these basic propositions: problems involving building up S&T capability, infrastructure and technology transfer, technological problems in the Caribbean, science and technology policies in Latin America, and the UNCSTD preparations.   The book ends with a presentation of a brief debate on the topics of future research on science and technology in Latin America and the Caribbean, and with a report of the Symposium.

Throughout the volume, one finds a single unifying element. All the contributors and participants seem to agree, either explicitly or implicitly, that the mobilization of scientific and technological potential for development is not an administrative but a political exercise that should involve not only scientists and technologists but politicians, educators, S&T users, and the public at large. The lack of concern of most of these groups in the LDCs with scientific and technological advancement of their countries, together with the work of the vested interests in the advanced countries, is largely responsible for the lack of progress in this area.

The convening of the Symposium would not have been possible

xii INTRODUCTION

without the financial support of Florida International University (through the cooperation of the International Affairs Center, the Caribbean/Latin American Studies Council, the College of Arts and Science, and the School of Technology); the University of the West Indies, and the University of Guyana. Dr. K. William Leffland, Dean of the International Affairs Center at Florida International University and Dr. Vaughan A. Lewis, Director, Institute of Social and Economic Research, University of the West Indies, deserve special thanks for their professional and personal support in convening the Symposium. Thanks are due also to Mr. Nicholas H. Morley for his financial support of work related to the publication of this volume, in addition to his major contributions to Florida International University as a member of the University Foundation Board of Trustees. Ms. Jane Marchi's insightful editorial assistance, Ms. Regina Greenstein's and Ms. Linda Goff's warm and cheerful assistance in typing the manuscript are greatly appreciated.

All the papers in this volume reflect the personal and professional views of their respective authors and do not necessarily reflect the policies or views of the organizations with which these individuals are affiliated.

I
Problems Involved in Building Science and Technology Capability

1 Building Scientific and Technological Capabilities in LDCs—A Survey of Some Economic Development Issues

D. Babatunde Thomas

One issue on which there is now general agreement is that scientific developments and technological advances were major forces in post-World War II economic growth throughout present-day industrially advanced countries (ACs). But the impact of these forces on the economic progress of individual, less developed countries (LDCs) has been either nil or very minimal, in spite of the political, economic, and, to a lesser extent, cultural "linkages" between most of these countries and a number of ACs.

The evidence suggests, additionally, that early exploitation of some nations by other nations, engendered by an ample supply of raw materials in the colonies of imperial powers, provided part of the basis for existing differences in national economic performances.(1) While the abridgement of the superficial linkages and dependence of LDCs on the advanced countries is vital to the attainment of self-determination and improved national economic performance by the former, the prevailing structure of the international system of trade and finance suggests that these linkages must necessarily be supplanted by mutually beneficial ones.

In the last two decades, the absolute and relative differences in the performances of the advanced countries vis-à-vis the LDCs have assumed major global importance for a number of reasons. The most critical of these is that the gap between the two groups of countries in terms of human welfare has been widening at an alarming rate. The continuing poverty and deprivation in the LDCs, and the relatively rapid economic progress in the advanced countries are manifestations of an ever widening gap. Although, during the same period, the rates of

3

economic progress in the advanced countries in general have been attenuated and subjected to periodic interruptions, the superficiality of the "linkages" and the dependence on these countries by the LDCs have tended to increase the adverse impact of such economic disturbances on the economies of the latter group of countries with greater severity than in the ACs.

The objectives here are to examine the fundamental question which arises from the foregoing background, survey some of the issues which must be understood in order to understand the underlying explanations for the slow rates of economic progress in LDCs, and discuss some ideas on what can be done to improve this state of affairs. The key question is:

<u>What are the circumstances and the limiting factors responsible for the inability of LDCs to benefit effectively from scientific developments and technological advances, and why?</u>

This question has been investigated repeatedly in studies on the development problems of LDCs, more so in the last two decades than at any other period in modern history, especially by social scientists and historians, and more recently by physical scientists and engineers. The preponderance of available evidence suggests that science and technology have contributed to incidences of <u>growth</u> in income and wealth in these countries, but have not contributed to <u>development</u> in terms of vital structural changes in their economies. A major, though simple, explanation for this situation is that LDCs lack the requisite scientific and technological (S&T) capabilities.

THE BASIS OF S&T CAPABILITY

The expectations of significant economic progress through science and technology, on the part of both international organizations and LDC policy makers and planners during these past two decades, have been based on a lack of proper understanding of the requirements for the successful application of modern science and technology to the problems of underdevelopment, and on distorted views of the impediments to the attainment of the longed-for progress. This lack of proper understanding originated from the tendency of policy makers and planners to treat science and technology as coterminous, or to focus primarily on the application of technology while sometimes inadvertently neglecting the development of science. This was largely due to ignorance of the proper link between science and technology as tools for development. Science may be defined as systematic knowledge of the physical and material world; and technology may be broadly defined as the vast sets of knowledge, experience, methods, and means of solving social and economic problems. A different problem exists in instances where the need for the orientation of pure and applied sciences to the building of S&T capability is recognized, for example, in parts of Latin America and Southeast Asia. There is a tendency in these situations not to recognize the building of indigenous scientific and technological capability as a long-term proposition. Thus, in both instances, the development of pure and applied scientific activities are not oriented to building the basis of S&T capability. Depending on the needs and the resource constraints facing individual countries or regions, this capability is composed of a predetermined quantity and quality of scientific and technological manpower to conduct pure and applied research and development activities necessary for national development, physical facilities (including laboratories and equipment), and other support facilities and services (such as educational institutions, libraries, consulting and engineering services, and technological information services). In sum, the basis of S&T capability is infrastructure.

In reference to the foregoing situations, the following questions may be posed and their relevance to the problems just outlined investigated.

1. To what extent, and in what manner, is science linked to technology?   or, specifically, is science the progenitor of technology?

2. Is there any evidence to support the proposition that scientific knowledge is not a precondition for the development and/or the acquisition and successful use of technological knowledge?

3. Are science and technology panaceas for social and economic development problems?

Although the approaches to these questions are not central to the primary focus of this chapter, certain aspects, which relate to possible responses by the international community to the scientific and technological needs of LDCs for development, are examined briefly.

It is often argued that LDCs do not need to duplicate the development of existing scientific knowledge as it is readily available through journals, books, symposia, conferences, and similar sources. Moreover, LDCs should not use their scarce resources to develop indigenous technology when the technology necessary for their development could be obtained through international transfer. It is further argued that the use of existing S&T knowledge is a short-term proposition and, through this approach, LDCs could avoid the inherent problems of trial and error involved in the development of new technologies, and thus minimize a significant part of the cost of developing their own scientific and technological resources.

There is no conclusive empirical evidence in support of these arguments. Although basically recommendations that are theoretically plausible, these arguments are ostensibly impractical. Neither the scientific nor the technological requirements of LDCs can be left entirely to foreign sources. Experiences of LDCs with public sector technical assistance programs, and their past dependence on international transfer of technology, suggest that this approach is neither feasible nor mutually beneficial to suppliers of the implied resources and the recipient LDCs. Furthermore, the international scientific, engineering, and technological literature constitutes a supplementary source of input into the building up of S&T capabilities. This source is readily accessible to LDCs, but its impact on the building up of the desired capabilities entails significant lead time.

Notwithstanding these considerations, the pursuit of problem-specific scientific research and the development of indigenous technology are necessary duplications and requirements for improved national economic performance. It is through such problem-specific activities that the extent of unique circumstances — for example, opportunities in agriculture for increased food production — could be ascertained and exploited. Science and technology, in the sense of pure and applied sciences, are interdependent. The distinction, and the suggestion that one can be pursued to the exclusion of the other, is mythical and has no basis in sound development policy. Granted that pure science is expensive, its selective use is essential to the unique circumstances in individual LDCs. Thus expenditure on pure scientific research can be limited to a fraction of expenditure on applied research and development.

The building of scientific and technological capability in LDCs must be understood as a means to specific ends. The ends are the enhancement of agricultural and industrial productivity, alleviation of poverty and deprivation, establishment of a relatively self-sustaining process of economic development, and a steady narrowing of the existing gap vis-à-vis the advanced countries. Furthermore, this capability must be improved continually in order optimally to derive the desired benefits.

In sum, the critical issues in building up viable scientific and technological capabilities in LDCs are, broadly, economic, social, political, cultural, and psychological. However, the economic development dimension of these issues is the most significant and most pervasive. The main economic development issues pertaining to building the necessary capabilities are infrastructural development which, when broadly defined, includes scientific, technological and managerial manpower development; as well as trade and the importation of foreign technology.

The following is a brief survey of these issues. The subject of trade is subsumed under the discussion of the importation of foreign technology. These issues are vital to the expansion of productive capacities in LDCs.

INFRASTRUCTURE

It is often argued that without adequate infrastructure or social overhead capital the process of economic development in LDCs is necessarily forestalled. In the broad sense, social overhead capital consists of transportation and communications systems; irrigation and drainage systems; public utilities; and health, education, and research institutions. All these services are basic to the functioning of a great many economic activities. In short, infrastructure comprises "those basic services without which primary, secondary, and tertiary productive activities cannot function."(2)

Typically, investment in the appropriate facilities and the services derived from them are characterized by external economies and technical indivisibilities. In view of these characteristics and the role of infrastructure in economic development, the necessary investment is considered in the national interest and the public sector generally assumes the responsibility of allocating investible resources for the creation of these forms of capital. There is some historical evidence of instances in which large foreign corporations have created some of these facilities for the operations of their businesses, on a limited basis, in LDCs where they were not publicly supplied.

The importance of infrastructure as a requisite for development is evident in the high percentage of total public investment accounted for by social overhead capital. In the last two decades, for example, the extent of the allocation of investible resources to infrastructure, as a percentage of total public investment, has been as high as 78 percent in Colombia; 41 percent in Ghana; 40 percent in India; and 38 percent in Mexico.(3)

Although the importance commonly attached to infrastructure is warranted, it ignores a critical factor — the building of a viable scientific and technological capability. This factor constitutes a vital dimension of broad infrastructural development. When this factor is not present, what exists is a less viable input for rapid economic progress. The services to be derived from basic infrastructure cannot be imported and, therefore, they must be provided largely through indigenous efforts. A major requirement in utilizing these services is the existence of scientific and technological capability.

The build up of S&T capability is not a requirement which must precede the creation of basic infrastructure; but it must, necessarily, parallel its development. The potential or resultant economic opportunities from the services of infrastructure can be exploited optimally only through the parallel build up of the necessary capabilities. This condition provides the basis for proper assessment and determination of the services needed for national development programs, and initiation of directly productive activities. This argument eschews the old controversy about the efficient sequence of investment in social overhead capital and directly productive activities. In the following analysis, the sequencing of these two forms of investment is of secondary importance because the existence of viable S&T capabilities provides the medium for the attainment of the desired goal, namely, the optimal exploitation of investment opportunities. In analyzing the critical role of S&T capability as a part of infrastructure creation for overall national economic development, the total lack, or poor knowledge, of viable private investment projects — projects of economic value — is a far more critical problem than the shortage of capital resources.

Existence of necessary infrastructure, in and of itself, neither reduces the cost nor enhances the profitability of directly productive activities.   By the same token, investment in directly productive activities are not necessarily autonomous or exogenous; rather, they are induced by knowledge provided through S&T capabilities, and a parallel build up of these capabilities along with the development of basic infrastructure. The linkage of social overhead capital with the building of S&T capabilities provide the basis for the productive sector to perceive the opportunities resulting from the creation of infrastructure. The ability to do so is vital also for the optimal exploitation of the social overhead capital, and for the effective and profitable expansion of productive capacities. It can be argued that the existence of excess capacity in social overhead capital in many LDCs (for example, underutilized port installations, periodically idle plants, etc.) is explainable by the absence, or the ineffectiveness, of linkages between the creation of infrastructure and the building of S&T capabilities. The extent to which these capabilities have been built up, and the extent of creative responses resulting therefrom, are functions of the quantity and the quality of the available scientific, technological, and managerial manpower. This human infrastructure is the basis of all S&T activities.

SCIENTIFIC AND TECHNOLOGICAL MANPOWER DEVELOPMENT

Historically, the experiences of the advanced countries and the LDCs have demonstrated the relative importance of skilled human capital resource formation in the development process. Natural resource endowment is a necessary, but not a sufficient, requirement for economic progress. Modern day examples of countries with little or few natural resources in the early stage of their development which have, nonetheless, achieved impressive economic progress include West Germany, Japan, Switzerland, Taiwan, and Singapore. Bearing in mind their respective forms of political and social systems, the experiences of these countries underscore, to a significant extent, the declining role of natural resources, and the increasing importance of human resources. The latter is the source of scientific developments and technological advancements which are potentially vital to the stimulation of economic progress and rapid alleviation of poverty. The relative importance of these characteristics is evident in the attempts to design and implement public policies in LDCs, specifically to increase the quantity and the quality of human resources.

The dismal results of LDCs' high dependence on importation of foreign technology also have given rise to varying degrees of commitment to the development and formation of the human resources necessary to build up viable local S&T capabilities. Science policy, technology policy, and education policy are all essential inputs into this endeavor. The focus of the last of these policies has to cohere with the first two to ensure the development of a sufficient number of adequately trained personnel, through time-phased teaching of science, engineering, technical, and managerial subjects in institutions of higher learning and, to the extent feasible, in elementary and secondary schools. In the latter case it is critical, for example, that the initiatives of the students be fully developed, and the conduct of science fairs encouraged.

The effectiveness of educational institutions, in terms of the education or knowledge imparted for the formation of human capital, depends upon the extent of initiative manifest, upon motivation, and upon the aptitudes and attitudes of those being trained. These institutions also have a major responsibility to promote appropriate attitudes for the acceptance of change. These are major considerations, which will tend to impede or facilitate the effectiveness of indigenous efforts in developing viable and ample reservoirs of scientific and technological manpower.

A new concept of education or training based on cost-effectness is vital to the proper delineation and integration of the activities of educational institutions in the building up of local S&T capability into the formulation of national science and technology policies. These activities must provide directions for the delineation of pure and applied research projects, which are consistent with the needs of the productive systems, and responsive to the implementation of national economic development programs. The training and nurturing of scientific and technological manpower is a long-term proposition, and, therefore, may not provide quick returns in response to short-term development plans. But the development of this manpower provides an important foundation for greater technological self-reliance in the long-run. A few examples of broad attempts at the formulation and implementation of S&T policies in LDC are the OAS Pilot Project on Technology Transfer and The Andean Pact (specifically decisions 24 and 84-89); and Canada's International Development Research Corporation (IDRC) project on Science and Technology Policy Instruments (STPI) involving eleven LDCs (six of which are in Latin America).(4)

The current trend in LDCs is towards the formalization of national policy on science and technology, and the integration of this policy into national economic development planning. The objective of the latter is the control of the composition and the direction of national economic activity, but this goal has rarely been attained. Apart from the apparent ineffectiveness of national economic planning bureaucracies, centralizing the building of scientific and technological capability through a comprehensive national development planning effort has its pitfalls. These are primarily institutional obstacles to S&T activities. The attendant centralization of power and decision making tends to constrain the development of new ideas and is often considered a high risk factor, real or perceived, on the part of foreign investors and potential suppliers of technology.

In numerous cases of the quest for cohesion, comprehensive national planning efforts have proven that there is a misunderstanding of this basic need to coordinate policies by government ministries directly responsible for the implementation of national development programs — primarily ministries of development, industry, and trade. Often these ministries do not have sufficient knowledge of what each actually does after the national plan document has been promulgated. Consequently, there is a tendency for them to function in mutually exclusive vacuums.

One approach to the minimization of institutional obstacles is to make existing institutions more effective, especially in the administration of S&T activities. This can be done primarily by promoting actives roles for scientist-administrators trained to provide appropriate context and framework for building up the desired S&T capability.

IMPORTATION OF FOREIGN TECHNOLOGY

The difference between the advanced industrial countries and the LDCs in terms of their respective rates of scientific and technological development is paralleled by the distinction between their trade in manufactured and primary products (agricultural crops and raw materials). The existing structure of world trade is one of the legacies of the colonial era. Although in certain instances the industrial countries are now buying more and more from the LDCs, the structure of trade has not undergone any significant or permanent change favorable to the LDCs in the last two decades. Between 1960 and 1975, the share of primary commodities in the exports of industrial countries fell from 48 to 24 percent while the share of manufactured commodities rose from 52 to 76 percent; imports of food fell from 17 to 11 percent. During the same period, the corresponding shifts in exports for LDCs were from 97 to 88 percent and from 3 to 12 percent respectively, and food imports rose from 16.5 to 17.5 percent.(5) It should be pointed out that a significant part of what is commonly classified as manufactured goods exports from most LDCs are products from assembly operations rather than from manufacturing in its entirety. This prevailing structure remains a formidable obstacle to the integration of LDCs into the mainstream of economic progress engendered by scientific developments and technological advances.

Furthermore, the role of international trade in the flow of technology from the ACs to the LDCs is anomalous in the sense that imported technologies, presumed to be largely embodied in capital equipment and machines, were never substantively internalized in the LDCs. Whatever positive contributions from imported technologies there are have been largely incidental in nature. The effects of these contributions have been demonstrated through growth in income as a result of transitory fortunes from foreign trade, and through channels of technology "transfers" in the form of direct investment, licensing, coproduction, turnkey plants with partial or complete product lines, management subcontracting, joint ventures, or technical assistance agreements. Although technologies flow through these channels from the industrial countries to the LDCs, experience suggests that these flows and the actual transfers have been spurious and disappointing, and have not made significant contributions to the economic development of these countries.

The primary factors explaining this situation are the character of the channels through which the presumed technologies flow to LDCs, and the non-fulfillment of the requirements for their adaptation, adoption, and application. Experiences in LDCs suggest that the 'transfer' of production technology by foreign firms through existing channels is neither targeted, nor intended to be internalized or indigenized in the host countries, further reducing their impact. More importantly, the reversal of this situation has been precluded by the absence of appropriate scientific and technological capabilities in the LDCs. One of the seeming paradoxes of the foregoing problems is that, in addition to wholly indigenous means, the importation of foreign technology is essential to the building of the desired capabilities. However, the process and the technologies imported must be selective and targeted.   The need for such an approach is central to the discussions of policy issues on "appropriate technology." Appropriateness of a technology to an LDC does not mean rudimentary or traditional technology. Appropriateness must be defined in terms of the needs and the circumstances of the country, industry, or firm for which it is intended. It preferably should be a production technology, capable of meeting the needs of the productive sector, and designed to utilize available local inputs and raw materials without, or with minimum, adaptation.   This technology must have a sound scientific basis, be inexpensive, effective, and capable of producing competitive products and services for the domestic as well as the world market.

The accessibility of knowledge from the international "technology shelf" is a necessary, but not sufficient, condition for the building of local capability.   In a recent paper, William Ellis argued that "the transfer of industrial technologies has prevented rather than assisted the economic development of the Third World."(6) The structure of the technology which is being introduced, and the process of the introduction, are influenced by input price distortions which tend to encourage inappropriate capital input bias (a scarce factor), which results in the transfer of inappropriate technology. It is doubtful, however, that the correction of these distortions and biases towards relatively scarce factors will necessarily encourage a bias towards relatively abundant factors. Out of the continuing discussions and analyses of these distortions, and approaches to their rectification, have emerged the argument that the flow of "appropriate technology" to LDCs needs to be encouraged. One of the basic challenges that·will face the proponents of appropriate technology is its viability in improving the competitive position of the users in the world market.

Attempts, through the public sector in the ACs, presumably to assist with the building of S&T capabilities in LDCs through technical assistance programs have not been as successful as might be expected. Two explanations for this result are the bureaucratic controls by the public sector in the affected LDCs, and the absence of a programmatic linkage of the assistance provided to explicit short- and long-term requirements in the productive sector.

Any study of the flow of technology from industrial countries to LDCs must take into account and examine the process of the flows, as well as the structure of the associated technological know-how. The structure of scientific knowledge and technological know-how may be devoid of normative matters; however, the process by which this knowledge is communicated from one location or one individual to another, and utilized, is highly subjective and dynamic. A plausible question now commonly asked apropos the issues facing the forthcoming United Nations Conference on Science and Technology for Development (UNCSTD) is the following: What can industrialized countries do explicitly to facilitate the flows of "appropriate technology" to LDCs, and, thereby, assist with the building of viable scientific and technological capabilities in these countries? Some of the proposed answers to this question are the subject of ongoing debates on UNCSTD; namely,

1. better terms − meaning lower cost of technology compared to past and current conditions;

2. less controls and few restrictions by suppliers of technologies, and

3. direct assistance with the building of local scientific and technological capabilities in LDCs.

These proposals, and the question to which they relate, are often misconstrued as suggesting that the industrialized countries take the initiative for the building of S&T capabilities in LDCs. But this has every appearance of the colonial past. It is, therefore, essential that the LDCs articulate their needs, and that the advanced countries understand them. Although the foregoing proposals are still lacking a specific program of implementation, they require the removal of undue external obstacles for the sharing of the industrial countries' scientific and technological power with the LDCs. It is this power that ensures the military, political, and economic dominance of the industrial countries. Any attempt at counteracting that dominance will invariably entail a restructuring of the existing international order. It is, therefore, understandable that the need for a new international economic order is currently a subject of heated debate between the LDCs and the technologically advanced countries.

One of the major considerations in this debate is the potential threat of confrontation manifested by social exigencies in the LDCs, the widening gap vis-à-vis the industrial countries, and the growing aversion to the sharing of scientific and technological power by the ACs. For example, labor unions in industrial countries, primarily in the U.S., oppose the purposeful export of technology to LDCs on the grounds that this exports employment. The fact that the technology flows can be mutually beneficial to the supplier and the recipient is one consideration usually ignored, probably because we still lack the necessary empirical evidence. Furthermore, there exist pervasive restrictive conditions and practices by suppliers of technology which mitigate against the adaptation and indigenization of imported technology. The prominent features of these practices are "inequitable" pricing of technology, secrecy to protect the "competitive edge" of suppliers, and limiting or prohibiting exports of output from the use of technologies supplied. All these constraints are subjects of immediate concern to LDCs. Existing schemes for the "pricing" of technologies have been criticized widely in LDCs as unfair and having no competitive basis. One approach to the promotion of "fair pricing" of technology is marginal cost pricing. But this subject is just one dimension of a larger consideration – an understanding of the structure and the process of producing technological knowledge (including the relative importance of research and development costs as well as improvement costs) in the measurement of the real cost of technology; the flow processes; and the utility of the technology to be acquired.

A SYNTHESIS

Fundamentally, there are three approaches to the building of S&T capabilities: 1) through indigenous (national) efforts;   2) through international efforts; and 3) through a synergy of both national and international efforts. The applicability of the third approach is ensured by the mutually nonexclusive character of the first two approaches. In terms of the feasibility and the practicality of the three approaches in LDCs, the building of S&T capabilities exclusively through dependence on international (private and/or public, bilateral and/or multilateral) efforts as part of a development strategy has to be viewed strictly as a theoretical possibility. Although an approach based solely on indigenous efforts is not totally useless, in some LDCs, dependence on it exclusively must be precluded in view of the urgent responses required by the development problems facing virtually all LDCs. The synergy of these two approaches, with variable emphasis on either of them as circumstances warrant, is the ideal strategy.

One of the conclusions in the preceding analysis of the importation of technology is that the existence of indigenous S&T capability is a precondition for the successful acquisition, adaptation, and application of imported scientific developments and technological advances. These complementary, rather than competitive, efforts are potentially capable of responding to the short-term, as well as the long-term development needs in LDCs. One offshoot of this complementarity in LDCs is the emergence and use of a mix of homemade technologies and foreign produced and controlled manufacturing technology, available through multinational corporations.(7)

A major obstacle to the success of the proposed approach is the choice of policy. The approach seemed to lend itself to an extension of import-substitution policy (for industrial development). This policy was pursued in many LDCs in the past two decades without regard to local unavailability of necessary inputs. Consequently, the policy was not effective in conserving foreign exchange nor in building technical capabilities.

In fostering the complementarity of indigenous and foreign technologies, and the substitution of the former for the latter over time, choices of foreign technological know-how and the processes for their acquisition and substitution in the LDCs must be relative rather than hierarchical. An important consideration in the resolution of these issues is that resources have to be channeled in such a manner that the flows are integrated with domestic efforts to increase the training of skilled (scientific, technological, and entrepreneurial) manpower, as well as the level of its aptitude, initiative, and ambition. Simultaneously, government incentives must be channeled to stimulate a complementarity of nonproprietary and proprietary technology. For example, a favorable legal environment and selective tax incentives are potentially useful in stimulating foreign suppliers of technology to remove restrictions and subtle constraints which have caused the suboptimal utilization of imported proprietary technology.

Through joint government-private sector R&D efforts, a viable approach could be designed to deal with the basic problems of small producers who lack the necessary resources for R&D activities, as well as to deal with the question of how to infuse technological know-how into the lagging sectors of the economy. Given the character of, and the production possibilities being exploited by, small producers in most LDCs, it is apparent that the preponderance of technology needed by the lagging sectors is of the nonproprietary variety. The urgency of the need for such technologies is usually not apparent in the activities of national research institutes in these countries. More fundamentally, the proposed approach could also serve as a vehicle for the general establishment of important linkages with the productive systems. This link is vital to the stimulation of demand for the services of scientific and technological research by those who need them; namely, the small producers and the lagging sectors of the economy in general. This must be the target group if poverty and deprivation are to be successfully alleviated. In order to beneficially utilize the nonproprietary or gratutitous technologies available in the advanced countries in the LDCs, attitudes towards their use must change. Nonproprietary technologies are commonly stigmatized as inferior and, therefore, are not being transferred or sought out by LDCs with the same vigorous demand as for proprietary technologies. Nonproprietary technologies are not subject to the same degree of controls as proprietary technologies, which by their very nature represent the state of the art. However, the requirement for beneficial use, in terms of scientific, engineering, and entrepreneurial foundations, apply equally to proprietary and nonproprietary technology.   The advantages of the latter are its relative suitability for the needs of small producers and lagging sectors, and its cost.

The increasing clamor for the involvement of the public sector in industrialized countries in promoting the flow of needed technologies to LDCs has to be treated with caution. Similar caution is necessary regarding the involvement of LDCs' national governments. In instances in which the governments have served as the chief technological "innovators" in LDCs, the results have been dismal. One explanation for this experience is the absence of the necessary connection with the productive sector, and a lack of profit-loss feedback on R&D activities.

One area which often is ignored or taken for granted is the development of managerial or "soft" technology. Without this form of technology, the successful application of production, or hard technologies, is doubtful. Soft technology constitutes the managerial, entrepreneurial, and organizational methods and experience responsible for efficiency in productive activities. While soft technology improves efficiency in production through improvement in organization and motivation, its association with hard technologies ensures expansion in productive capacity and growth in productivity. The effectiveness of soft technology, and, therefore, efficiency and productivity, are means to an end. The end is the resultant improvement in the standard of living through, for example, increased supply of food and shelter together with the means to consume them, and the general enhancement of the quality of life of the masses.

The primary factor in international efforts to build viable scientific and technological capability in LDCs is the private sector in industrialized countries. The counterpart of this factor in the LDCs consists of the creation of infrastructure, especially scientific, engineering, technological, and entrepreneurial (STE and E) manpower and appropriate infrastructural support. The role of the public sector in general must be to provide incentives, and avoid wholesale superficial regulations; and where active involvement is deemed in the national interest, such cases should be selective, and partnership with the private sector encouraged.

Finally, the flow of imported scientific developments and technological know-how must be monitored by an independent body composed of the STE and E community to ensure that their configuration and rate of flow can be productively used locally, both to meet immediate needs and for the development of the capacity to meet future needs, and in shaping a positive social impact for science and technology.

One general conclusion drawn from the analyses of the key question raised here is that its basis is fundamentally economic, but the rudiments of the solutions are manifestly political. It is by acknowledging this basic conclusion that progress could be made towards an understanding of scientific and technological issues in the development of LDCs.

NOTES

(1) The following critique by Kwame Nkrumah comes from one of the strongest and most forthright expositions on these differences. ". . .In the last quarter of the nineteenth century, colonies had become a necessary appendage for European capitalism, which had by then reached the stage of industrial and financial monopoly, that needed territorial expansion to provide spheres for capital investment, sources of raw materials, and strategic points of imperial defense. Thus all the imperialists, without exception, evolved the means, their colonial policies, to satisfy the ends, the exploitation of the subject territories for the aggrandizement of the metropolitan countries." Kwame Nkrumah, Africa Must Unite (New York: International Publishers Co. Inc., 1970), p. xiii.

(2) Albert O. Hirschman, The Strategy of Economic Development (New Haven: Yale University Press, 1967), p. 83.

(3) Henry J. Bruton, "Social Overhead Capital," IESS, (New York: Crowell, Collier & Macmillan, 1968), p. 288.

(4) See respectively, F. Sagasti, A Systems Approach to Science and Technology Policy Making and Planning, (Washington, D.C.: Department of Scientific Affairs, OAS, 1972), and STPI Project, A Comparative Research Effort to Examine Ways and Means of Implementing Science and Technology Policies in the Industrial Sector, (Lima, Peru: Office of the Field Coordinator, 1974).

(5) See IBRD, World Development Report 1978, Washington, D.C., 1978.

(6) William N. Ellis, "A.T.: The Quiet Revolution," Bulletin of the Atomic Scientists, November 1977, pp. 24-29.

(7) Miguel Wionczek, "Less Developed Countries and Multinational Corporations: The Conflict about Technology Transfer and the Major Negotiable Issues," December 1977 (mimeograph).

REFERENCES

Bauer, Peter T. Dissent on Development: Studies and Debates in Development Economics. London: Widenfeld and Nicolson, 1971.

Hirschman, Albert O. The Strategy of Economic Development. New Haven: Yale University Press, 1967.

Lewis, W. Arthur. Some Aspects of Economic Development. Tema, Ghana: University of Ghana and Ghana Publishing Corporation, 1969.

Nurkse, Ragnar. Problems of Capital Formation in Underdeveloped Countries and Patterns of Trade and Development. Oxford: Oxford University Press, 1967.

Thomas, D. Babatunde. "Obstacles to the Transfer and Adaptation of Imported Technology in African Countries" in Harold Davidson et al., eds., Technology Transfer. Leiden: Noordhoff International Publishing, 1974.

_____. "La transferencia international de technologia industrial y las naciones nuevas," El Trimestre Economico 41, no. 3 (July-September 1974): 605-624.

_____. Capital Accumulation and Technology Transfer. New York: Praeger, 1975.

Wionczek, Miguel S. "Some Questions for the World Jamboree," Bulletin of the Atomic Scientists, December 1977, pp. 29-32.

2 Building National Scientific and Technological Research Capability in the Context of Underdevelopment

R.O.B. Wijesekera

In discussing national scientific and technological capability in the context of underdevelopment, I would like to begin with a series of definitions because the words "science" and "technology" do not always mean the same things to scientists, nonscientists and those in-between.

I consider science as the activity that generates new knowledge about the nature of things and about phenomena. This activity is characterized by a sequential methodology known as the scientific method that comprises observation and development of ideas, experimentation, hypothesizing and verification of hypotheses by further experimentation leading to the theories which are the basis of scientific knowledge. These theories are periodically subjected to scrutiny by the same sequence and form a body of knowledge which keeps changing and expanding, as new theories and new experimental techniques emerge to replace the previous ones.

Technology is the outcome of the attempt to use science and scientific knowledge for human needs — be they good or evil. Therefore, inasmuch as science creates a body of knowledge, technology gives rise to new processes, gadgetry, machines, devices, and instrumentation. So immediately one sees the most important distinction between science and technology, namely that the outcome of technology is more readily "visible" than the outcome of science.

It should be kept in mind that during the course of history, technology preceded science. Man, by trial and error methods, experimentation, and innovation, developed certain empirical technologies that were not based on a body of knowledge accumulated via anything like the scientific method I have described.

In contrast to these empirical technologies of other eras, modern technology is comprehensively based on science, and must, therefore, maintain continued links with new scientific knowledge in order to sustain its own advancement and to serve its special purposes — the expansion in production of goods and services and the improvement of the quality of life, the elements of the modern concept of development.

17

CHARACTERISTICS OF UNDERDEVELOPMENT:
THE LACK OF A CULTURE SUPPORTIVE OF SCIENCE

There are many characteristics that help to identify the "under-developed" countries: comparatively small populations; prescience cultural heritages and colonial political backgrounds; underutilized and underexplored natural resources; unemployment and scarcity of food; balance-of-payments problems; political chauvinism; lack of scientific and technological capability; absence of public demand, understanding, or sponsorship for science; prevalence of outmoded bureaucracies; lack of leadership in scientific research or managerial capability; and overdeveloped trade unionism of Western type.

Many of the underdeveloped countries live in the world of prescience cultures(1) and represent tradition-bound societies that were built long before the advent of the industrial revolution. The nations of Asia and Africa, many of them − India and China excepted − with populations between 10 and 50 million, can indeed be characterized as prescience cultures. They knew no science; although admittedly, they knew philosophy and displayed high levels of empirical technology.

By the time of the industrial revolution in Europe, many of these countries had already come under colonial subjugation and remained subject states until the post-World War II era. When they emerged as sovereign states committed to modern development, science and technology had already been recognized in the world as the dominant force. Their own "science," however, which may be termed colonial science − for want of a better term − was limited to the barest minimum. Most scientific activities conducted within these countries were carried out by scientists linked to the colonial powers, and were related to their objectives such as the exploitation of the natural wealth of the colonies. Consequently, there was, in some cases, a certain measure of research in forestry, agriculture, and mineral extraction of the former colonies.(2) Scientific research on tropical diseases also falls into this category.   Thus, the "colonial science" was largely of an applied nature relying very heavily on the fundamental base of research carried out in the colonial mother-country. The development of an indigenous scientific tradition or scientific manpower at that stage would not have been possible because of the unsurmountable problems created by traditional attitudes, the lack of the infrastructure, and even of the understanding of the nature of, and need for, science.

This situation has accentuated the feeling within many social subjugated peoples that science was an alien exercise and could never be carried out within their country. Planners in many underdeveloped, but potentially rich, countries still have doubts about their ability to build their own scientific community and traditions. In some instances, this insecurity translates itself into chauvinistic attitudes that manifest themselves in an aggressive movement to revert to prescientific tradition and deny the possibility to use science to change the society. These attitudes say "This has been done here for 2,000 years, what can 'science' do for it now?" These negative sentiments, reinforced by the attitudes towards the world and life taken from Rousseau, are as tragic as they are untenable.

Not all the attitudes toward science, technology, and development that characterize the developing countries are of ancient vintage and the direct result of colonialism. For instance, in many developing nations there has been (and still is) a tendency to slavishly copy the policies that the more developed nations have used in their own very different context. Thus, the period 1950-65 can be categorized as the age of "development by economists." It was fashionable in many underdeveloped nations to draft in that period five- or ten-year development plans. These economic plans failed to realize the need to build a scientific and technological structure that would help the development process. Very often the existence of such structures was implicitly accepted where none actually existed. The fact which was not realized was that many of the developed countries, from which the techniques of drafting these plans had been borrowed, had well-established scientific and technological structures. The economic planners thus assumed the existence of such plans in LDCs.

The adoption of a more realistic approach by developing nations in the 1950s and the 1960s would have enabled them to realize that they faced the great need to include in their development plans the objective of building up of scientific and technological manpower, for both long-term and short-term goals. (India was one of the few countries that became aware of such necessity.) One of the manifestations of the economic plans of the 1950-65 period is the presence today in many developing countries of ill-conceived, inappropriately designed industrial ventures, purchased on a turnkey basis and incurring heavy financial losses. The industrial sector, generally, suffers from both management and technological deficiencies and carries the burden of a lack of the local research and development competence. Thus, the industry displays the technological features imposed from the outside upon the countries devoid of any scientific and technological capability that would permit them to assess their own needs. Consequently, the industrial sector is plagued with process and machinery obsolescence, excess capacity, products of substandard quality, and the inability to sell abroad.

The countries that have thus embarked on such industrialization programs divorced from scientific and technological advancement find themselves grappling with the problems of balance-of-payments, of marketing their industrial production, of labor unrest, unemployment, and even food scarcity. The total economic and social price for industrialization, that did not give any recognition for the need to develop a scientific and technological base, was quite high.

In many countries, ideological considerations and political chauvinism have dictated that the industrialization programs should be government controlled. Such decisions, inspired by the examples of the socialist countries of Eastern Europe, though understandable, overlook several factors. First, in many developing countries there is hardly a scientific tradition or a technological expertise comparable with that available in the socialist East European countries. Second, the developing countries fall far behind Eastern Europe in respect to managerial competence. Third, many developing countries are saddled with inflexible and outmoded bureaucracies which, acting as the managerial arm of government, are committed to run state-owned enterprises.

These factors plus the outmoded concepts of "public accountability" virtually doomed the public sector industries to nonprofitability and waste. The problems of the state enterprises were further aggravated by the excesses of trade unions slavishly affiliated to leading political parties. In the countries with two main political parties – following the British model – the sharply divided trade unions can act in a manner detrimental to the industry in order to deliberately undermine a government and score a political point. The labor's ad hoc demands take priority over rational industrial planning.

Furthermore, the siting of new industrial enterprises becomes a political exercise, in which political considerations hold sway and technological requirements are ignored. While the industrial decision makers are sometimes unaware of the consequences of their actions, a generally servile bureaucracy concurs with the political choices. The opinions of scientists or technologists matter very little.

It is in such a milieu – a climate hostile to science – but a national science and technology capability has to be built, nurtured, and maintained in many developing countries. The small national scientific communities are overburdened with teaching in higher educational institutions, research, and servicing the public sector. Not the least of their heavy burden is the obligation to participate in a plethora of meetings, seminars, discussions, and consultations. On the other hand, the public awareness of the need for science and technology in development, and still less the public sponsorship of science, are absent. In many instances almost all the science is carried out in governmental or government-sponsored institutions, and its sole funding source is government. Bilateral aid or loans from international sources are seldom available for activities directed towards building a scientific and technological manpower and, thereby, a national S&T capability.

Thus, the small struggling scientific communities in many developing nations not only lack, as a rule, any sort of political patronage or recognition, but are faced with other disadvantages. Infrastructural shortcomings commence with the restricted nature of the educational base from which potential S&T manpower must spring. The short-comings of technical training, the lack of scientific information, and the restrictions imposed upon interaction with the international scientific community work against any research activity.

REQUIREMENTS FOR THE SUCCESSFUL GROWTH OF SCIENTIFIC ACTIVITY

Among the requirements for scientific advancement in the LDCs, nothing is more important than a climate "supportive of science." Its main components are:

1. an educational structure geared to produce scientific and technological manpower required to sustain an implantation of science and to encourage the emergence of scientific leaders;

2. political patronage and adequate domestic expenditure to ensure that scientists and technologists are sufficiently motivated to do creative work;

3. measures to enable the national scientific community to operate as part of the world scientific community.

4. recognition of merit as the criterion in the selection of scientific leaders and advisors in science and technology matters; and

5. a national science policy body with the correct perspectives and sensitivity to serve the needs of the scientific community.

The Educational Structure

Following their political decolonization many underdeveloped nations started restructuring their educational systems. In several countries, like Sri Lanka, state-sponsored schemes of free education were introduced. The philosophy underlying the free educational scheme was a laudable one – to extend the benefits of education to the entire population, both rural and urban. The rapid creation of a large number of schools in new areas led immediately, however, to shortages of teachers, equipment, books, etc., and a consequent lowering of schooling standards. While teaching science was emphasized, and the status of the schools was frequently judged by the presence and the absence of "science teaching facilities," the results were far from satisfactory.

The new system tended to over-simplify science, de-emphasize its international character, and obscured the nature of scientific principles in favor of the mechanical teaching of "scientific facts." Although this helps in diffusing factual scientific knowledge throughout the population, its very egalitarianism mitigates against the emergence of those with the creativity and aptitude to develop as scientific leaders. The very similar situation in other Asian countries can be explained partly by the language problems of the region. The adoption of local languages for nationalist reasons at the lower and middle education level created obstacles to university education in the sciences, research activity, and the development of a national scientific capability. Since education reforms in the Caribbean are still at an initial stage, there appears to be much to be gained by examining some of the experiences of the smaller nations of Asia, prior to any schemes being put into operation. While the preindependence situation in the Caribbean was similar to that which prevailed in the former British territories in Asia after the last war, the British Caribbean is, fortunately, free of the complications arising from the diversity of indigenous languages.

If a scientific and technological capability is to be built, new educational policies must be oriented from the primary school stage to develop the attributes necessary to produce scientists, technologists, and technicians, such as natural curiosity, powers of observation, and capacity to reason and innovate.

The manpower resources necessary to acquire a minimum science and technological capability consist of:

- scientists and technologists for research and its support;

- technical middle-level manpower comprising assistants, machinists, maintenance and service personnel, skilled operators, etc.;

- conceptual and management supportive manpower comprising science and technology policy makers and managerial staff for R&D activities; and

- specialists in scientific and technological information and documentation.

To achieve such adequacy, the entire educational system, from the primary school to the institutions of higher education, must receive – in the language of the space age – "a fresh engine thrust and a course correction."

While several developing countries have now programed the overall restructuring of their educational systems, a serious weakness can be observed in these reforms. While much funds and effort are channelled into the educational system, and sometimes even to the universities, little, if anything, is done in respect to expenditure supporting the scientists and technologists engaged in research effort. While, according to UNESCO, any country should dedicate one percent of the GNP for investment in R&D, there are many countries in the region that do not spend even a quarter of this amount on R&D. Even when funds are available, only a fraction of the money appropriated for research is actually expended, reflecting one of the paradoxes of underdevelopment – even the scarce available manpower is not used rationally and productively.

Patronage for Science and Meaningful Investment in Basic Science

High-level political patronage for science was an important factor in the development of science itself in today's advanced Western countries. Moreover, the main motivating force for the development of Indian science in the decades following World War II, to its present enviable state, derives from the ponderous patronage given it by Jawarhalal Nehru.

Political patronage for science is even more important in the context of today's underdevelopment than it was in the eighteenth and nineteenth century West because neither can a scientific and technological competence be built in a developing country without active political support, nor can an educational system needed for scientific advancement be constructed without political decisions.

Since the private sector in developing countries does not actively promote S&T activities as its counterpart in the advance countries, the development of the national capability in that field in the LDCs rests largely on the patronage and financial support of governments. However, governments of developing nations do not always realize that it is clearly in their interests to support science and technology, or they accept incorrect attitudes toward S&T advancement. Those charged with decision making in that field are nonscientists or former scientists with comparatively poor career records who are overly impressed by literature and fashions coming from the developed countries. Once in bureaucratic positions, these people often argue that the only thing worth being supported by the government is what they term "applied research" or "technology."

It is almost impossible to justify basic research to nonscientists anyplace, but the task is even harder in the LDCs, where — in spite of supportive experience — no people in high places understand that technology and applied research can never survive as a self-generating activity without the necessary component of basic research.(3) As an outstanding Indian scientist, Nayudamma, former head of the Council of Scientific and Industrial Research in his country, has stated:(4)

No country can prosper simply by the importation of research results. Every country must do research on its raw materials and its natural resources. It must form and maintain its own scientific personnel and it must develop its own scientific community.

While in the developed countries the futile debate of "applied science" vs "pure science" takes place, similar debates in the LDCs are meaningless and disastrous. Since the developed countries have already built up their research capabilities, they may consider how to use that capability to serve short-run needs. In the LDCs, however, such debates demoralize an embryonic scientific community struggling to emerge and survive. The overemphasis on the so-called "applied research," which in the context of underdevelopment is most often simply extension work or copying work done elsewhere, opens the door to bad and irrelevant pseudo-science and drives serious scientists abroad. Furthermore, the range of choice available from among the alternative sources of foreign technology are severely constrained because the exercise of such choice assumes the availability of a degree of knowledge and sophistication which, in turn, depends upon the indigenous science and technology infrastructure. For example, there is a wealth of information collected from satellite data on forestry and marine resources, geology and climate of many developing countries. However, such information will only be of value to the LDC if it has capacity for transformation of the information from computer tapes to maps, tables, and other readily usable forms. This means that a modern scientific and technological capability must include such things as advanced computer science, and technology needed to make use of available information.

Benefits accruing to an indigenous scientific community from interaction with international science exceed the assimilation of knowledge. Such interaction removes the feeling of isolation and impotence prevalent among LDC scientists.

Partnership in World Science

Science, by its very nature, is international, and any scientific community must recognize the universality of the standards of science. As an outstanding United States chemist put it:(5)

> In scientific research there exists only one standard of excellence, namely an international one. A hypothetical statement such as 'this is a very good chemical research for Kenya but rather poor for Sweden,' is equivalent to saying: poor chemical research is being performed in Kenya.

The international standards for science also mean international recognition, if these standards are met. A scientific community which has developed to a level where it has the capacity to produce good scientific work, even in limited areas, has also, thereby, developed the capacity to participate fully in the world scientific advancement in that area.

If these simple facts are not recognized, a country is committed to the necessity of having to import most, if not all, of scientific knowledge, and having to rely on foreign technology – often executed by foreign personnel.

International science has its own complex structure, and every scientific community, however small, must have the opportunity to operate as a part of this community. This means participating in world meetings, seminars, and refresher courses, and even working periodically in internationally reputed research centers. All these activities must be recognized as an integral feature of science and technology advancement by S&T decision makers in the LDCs. Unfortunately, the harassments suffered by scientists in some LDCs, when they aspire to take part in international scientific events, are occurring every day.

Scientific and Technological Leadership and the Recognition of Merit

The distinction can be made among several levels of scientific leadership. There are leaders of scientific and technological groups involved in research on a specific problem, or committed to the completion of a specified project; leaders of divisions or departments of research centers that direct and oversee groups of research projects; and leaders of institutions who are charged with broader policy decisions.

In the developed countries the state of scientific advancement and the existence of a strong scientific community facilitate the emergence of competent leadership. In the LDCs, the experience has been most unfortunate. The lack of proven and experienced research leadership at all levels results in the use of suitably recruited expatriate advisory personnel or the employment of apparently the best available local staff. The indiscriminate use of local scientists, which may reflect short-sighted nationalistic considerations, may have quite disastrous results, because it may mean that third-rate men presently available would have charge of the destinies of potential first-rate scientists.

A commendable, and perhaps most reasonable, policy would be to identify potential young leaders in the lower levels of the scientific establishment and expose them gradually to the problems posed by leadership at the highest level. The least could be expected, on the other hand, from the employment of local exscientists and administrative personnel as scientific leaders. This procedure, so frequently observed in developing countries, has resulted in unbelievable malpractices. The administrative and managerial "leaders" are prone to overemphasize the need for administrative personnel, demoralize the researchers, and shift the work towards pseudo-science which falls within the superficial range of their comprehension. In the case of technology, such "leaders" tend to accentuate what in some developing countries is known by scientists and technologists in mock scorn as "political technology."

The shortage of a competent and experienced scientific and technological leadership at the highest level is not the most unfortunate aspect of underdevelopment. The build-up of other categories of scientific leaders is perhaps more vital since they will represent scientific and technological capability in the future. The correct and timely recognition of potential leadership is, therefore, of paramount importance.

Though difficult to assess, scientific and technological merit, judged, if necessary, with assistance from scientific and technological personnel from outside, must be the sole criterion of appointments, promotions, and recognition within the S&T community. In developing countries in Asia (and maybe also in the Caribbean region), the scientific and technological performance and intellectual quality of the heads of research institutions, university professors, and heads of scientific departments, even vis-à-vis their own subordinates, leaves much to be desired.

Political reasons apart, merit is not recognized in the underdeveloped S&T communities, and nothing is more detrimental and demoralizing to good young scientists which many countries possess than the apparent recognition offered to stooges, charlatans, and manipulators. At the policy level, the situation is similar: decision-making politicians in the LDCs have to contend often with S&T advisors of doubtful quality or who lack the appropriate experience.

National Science Policies and Policy-Making Bodies

In the early 1960s there was strong pressure in many developing nations for the formulation of national science policies. What was, perhaps, not realized then, although scientists assumed it, was that science policy in the embryonic stage is only meaningful when it concentrates upon the development of scientific and technological capability. The other aspect of science policy is the utilization of that capability to serve the needs of national development. Today, many developing countries in the world possess national science policy organs − the UNESCO directories list a great many − but scientists are not convinced that the establishment of these agencies has done much to enhance national scientific activity or to promote technological adaptation and innovation. In many instances, the national S&T policy-making organs have been remote from scientific and technological activities and were not provided with power or resources to make an impact.

Scientific and technological development in a LDC has to contend with: 1) a set of immediate requirements such as health, housing, and food, important for both political and humanitarian reasons; 2) the longer-term needs of development of natural resources, e.g. mining, agriculture, hydropower and irrigation, forestry, etc.; and 3) planning of future infrastructural requirements.

While it is natural and expedient from a political viewpoint to put emphasis on the short-term needs, the long-term needs move up all the time to become short-term and reach crisis dimensions. In order to avoid such crisis situations, it is imperative that all science and technology programs take a balanced view of short-term and long-term objectives.

To a crucial question: Where is the capacity to do all this, if a scientific and technological capability does not exist?, one may answer with the following assertion:

> The requirements for science policy formulation and for technol-ogy analysis of each country are unique; they depend on its development objectives, its size, its resource potential, and its capabilities in Science and Technology.(6)

Therefore, for national science and technology policy organs to succeed in the developing nations, they must first be linked to the highest political authority. Secondly, they must have command of the appropriate resources, and third, they must be free of exceedingly nationalistic attitudes that inhibit utilization of outside personnel for advice when no competent local scientists and technologists are available. Until these conditions are fulfilled, the purpose of national S&T policy agencies would remain largely ornamental, and they will be a symbol of a self-imposed, but irrelevant, sophistication.

MEETING THE ISSUES

The realization that scientific and technological advancement depends upon favorable political decisions is just the first, but also the decisive, step toward the establishment of S&T capability. Many such steps, in several simultaneous directions, are necessary before a scientific and technological community can be built and molded to serve the purpose of national development. The policy components involved fall into two broad categories: those that can be put into operation by

autonomous domestic decisions; and those that involve external assistance and interaction and, therefore, external decisions as well.

As the role of education at all levels in the development of
scientific manpower has been given attention earlier, some remarks are
now in order about the importance of R&D programs for the building up
of a level of competence in science and technology. Every country has
some nucleus of trained personnel. To engage this nucleus into relevant
research and maintain it productively is of as much consequence as the
long-term programs for manpower building. Towards this end, the level,
character, and productivity of R&D programs, the manner in which they
enhance efforts in industry, health, agriculture, forestry, and public
utilities, bear much significance.

Scientific and technological manpower thrives on creative opportunities, and, if these opportunities are provided by state agencies in the
LDCs, then in order to display social visibility research activity must in
some way be associated with the major development areas. Given that
basic research is a necessary ingredient to sustain both applied research
and technology, basic research in a developing country also must be
built around these major subject areas. For example, some basic
research in chemistry, engineering, and biology can easily be built
around research in agriculture; research in genetics, entomology,
biochemistry, pathology, physical and inorganic chemistry can greatly
enhance productivity and relevance of specific applied agricultural
research.

The important issue, therefore, is that the basic research should as
much as possible be associated with ongoing applied research and/or
technology. Such an approach will create manpower resources related
by a common broad subject area on the one hand, and with a vertical
integration from basic research to technology on the other. In this way,
interest in all segments of research will be fostered, and a sense of
involvement is more likely to arise in the S&T community.

Institution Building, Research Groups, and Infrastructural Problems

In many developing countries R&D institutions tend to be rigidly
structured. The nature of the service conditions that bind the personnel
and the bureaucratic principles involved in their operations are not
helpful to the formation of permanent research groups that are a very
necessary feature of well-organized and productive research. It is not
easy to define what is meant by a research group:(7)

> In most cases the research groups constitute an informal level of
> organization and as such are seldom created to conform to some
> standard design. They can assume a wide variety of forms
> ranging from the strongly integrated local research teams
> operating over long periods of time and working on a single
> precisely defined objective, to the very loose and transitory
> collaborations among scientific peers working at a long distance

from one another. Given such variety it is not easy to propose a single precise definition of the research group.

The following definition appears, however, adequate:

A research group is a group of scientists (at least two) working under common intellectual leadership in a well-defined research specialty. The members of the group engage in continuous or periodic research collaboration. The leader is an active researcher in the group. A scientist may belong to more than one research group. The members of the group may be recruited from two or more formal units/organizations.

Throughout the history of science the occurrence of the research group feature can be identified, and today it is customary for researchers to refer to work as performed by so-and-so's group. The importance of research groups in enhancing productivity cannot be overstressed. The development of clusters of strong research groups in selected areas of activity is the best method of building up science in the smaller developing nations.

A major structural change in formal institutions which allow and indeed activate the process of formation of research groups in developing countries must, therefore, be promoted. The important aspect of research group formation is that the group leader is the accepted choice of all his colleagues. The fate and progress of the group, and thus the career prospects of each constituent member, is comprehensively dependent on the group's collective productivity; and leadership is a major factor in this regard.

The formation of research groups will, in time, tend to override interinstitutional petty rivalries and frictions between research institutions, government departments, and university departments, which frequently render collaborative research a nonstarter in many developing countries. Research groups make it easier to adopt the multidisciplinary approach, often so very necessary in solving problems in the underdeveloped context.

International Interaction and Assistance

Any attempt to develop a domestic scientific and technological capability would be doomed without mechanisms for interaction with the world scientific community. Three aspects of international interaction may be identified broadly: acquisition of available knowledge, synergistic responses on indigenous research, and initiatives in collaborative research.

There are many formal mechanisms available today for the acquisition of knowledge in science and technology. However, since the depth and extent of its assimilation is dependent on the level of sophistication, there is need to develop active indigenous research activity in an underdeveloped situation if full use is to be obtained from the world's existing knowledge.

Synergistic responses to indigenous research generally stem from a scientist-to-scientist interaction. Scientists working in the same field correspond with one another and form what are now known as "the invisible colleges" of science. This type of contact is most prevalent, even in the developing nations, and in many a case is responsible for sustenance of research activity under the most difficult conditions. Personal contacts, afforded by scientific meetings, enhance this type of interaction. Frequently, they represent a donor-receptor situation, where a scientist from a developing country benefits from the association with a scientist from a developed one. Occasionally, when interests of the two scientists complement each other, it is a two-way benefit. The linkage, complementary between the leaders of two research groups, helps to enhance productivity and capacity of the less developed group. Formalization of such a model, based on the "adoption" (in a scientific sense) of a group in an underdeveloped situation by a group in a developed country with similar research interests, has been proposed and commended to international agencies.(8)

Initiatives in collaborative research have had some success but have always depended on too many extraneous bureaucratic, political, economic, and geographic factors that inhibit scientific growth in the LDCs.(9) In addition, several other mechanisms, like bilateral cooperation, centers of excellence, and regional networks, have their merits as well as their deficiencies. A discussion of these forms of cooperation falls outside the scope of this chapter. But it should be stressed that regional mechanisms do not dispense with the need to develop indigenous scientific and technological capabilities in most LDCs.

NOTES

(1) Stevan Dedijer "Underdeveloped Science in Underdeveloped Countries," Minerva, (1963), 62.

(2) R.O.B. Wijesekera, "Science in a Small Developing Nation," Scientific World, (1976), 1:6.

(3) R.O.B. Wijesekera, "Enlarging of the Bounds," (Third Annual Lecture of the National Science Research Council of Guyana, June 1977, Georgetown, Guyana.)

(4) Y. Nayudamma, "Promoting the Industrial Application of Research in an Underdeveloped Country." Minerva 5, no. 3 (1967): 323.

(5) Carl Djerasci, "High Priority: Research Centres in Developing Nations," Bulletin of Atomic Scientists, 1968, No. 24, p. 31.

(6) F.A. Long, "Effective Organization and Modification of Technology for Development," Research Policy Program, Cornell University, Ithaca, N.Y., 1977.

(7) R. Stankiewiez, "Research Groups and the Academic Research Organization," Sociologisk Forskning 13, no. 2 (1976): 20.

(8) R.O.B. Wijesekera, A New Mechanisms for Developing Scientific Research in the Developing Nations – the Group Adoption Model. Proceedings of 25th Pugwash Conference, Madras, India, Jan. 1976; R.O.B. Wijesekera and Marina Wijesekera, Proceedings; Special International Sessions of the Sri Lanka Association for the Advancement of Science. Colombo, Sri Lanka, July 1976. See also note 1.

(9) M.J. Moravcsik, Science Development: Towards the Building of Science in the Less Developed Countries. International Development Research Center, Indiana, 1975.

3 Science, Technology, and Education— The Problem of Human Resources

Rodrigo Zeledon

Nowadays we take for granted that science and its applications are mandatory for the cultural and material development of a country. We also take for granted that every country needs some critical number of scientists of high quality who must raise the local scientific level, and work directly and/or indirectly toward the solution of national problems.

These premises are as valid for less developed countries as for the more developed countries. However, if these premises are not observed, LDC development will not be able to count on the indigenous component necessary to free them from their nearly total technological dependence, and consequently to work toward a rational equilibrium of their precarious, and often unfavorable, balance-of-payments.

GENERAL BACKGROUND

The Deficit of Scientists

To me, development denotes the well-being of the majority with respect to health, nutrition, housing, education, recreation, and all the other minimum material and spiritual conditions that modern man requires. So defined, it sounds trite and easy enough to achieve in practice. However, it has been an almost unattainable goal for most Latin American and Caribbean countries. There is an evident divergence between the statements of politicians, planners, and national leaders as to the achievements and potential value of modern science, and their actions to utilize these advantages in order to create a sound, scientific tradition.

In spite of the positive steps that some countries have taken in recent years, the truth is that the number of scientists and specialists is still regrettably low in the area. Statistics show that in advanced countries the scientist/inhabitant ratio is 1 to 1,000 or more, while in

31

the LDCs the ratio is significantly lower. Nevertheless, it is my opinion that the data on LDCs have been gathered in a none too reliable fashion. If we are to apply a rigorous scientific criterion, such as that used for advanced countries, as to who should be considered active scientists, we would be alarmed to see that the number of active scientists is even fewer, and the quality of our research and of our scientists is, in general, inferior to what we have been led to believe.

It is also evident that budget allocations for research are minimal, in proportion to the national gross product, as compared with the amount of money invested in countries where vigorous scientific development is regarded as a prime requirement for the attainment of material and cultural well-being.

Development of an Indigenous Science

What are the factors responsible for the slow growth of indigenous science in our countries? The problem is complex and no easy answers nor solutions exist. Deep-rooted historic and sociopsychological factors and, more recently, sociopolitical factors also have been instrumental in determining the generally not too promising scientific profile seen in most countries of our region.   M. Roche suggests that, among other factors, deep-seated religious prejudices have been strongly detrimental to the establishment of an adequate "scientific climate" in the majority of Latin American countries.(1)

There is a popular defeatist opinion that research carried out by local talent has no place in produce-important material, intellectual, and cultural dividends.

Sound scientific planning based on the reality in each country and with the concurrence of local scientists, while respecting individual creative freedom, is undoubtedly the formula that leaders responsible for the planning of science in each country should pursue. While we believe there is no universally acceptable formula, since scientific policies must rise from the realities of each country, the tenets implied in a policy of "systemism," as Bunge propounded, are pertinent and to the point.(2) Kalnay presents a similar view of scientific planning, stressing the point that each country must be scientifically prepared to solve new or unexpected problems, not just those in a given historical moment of its development.(3)

This leads ·to another important aspect in the development of science and the human scientific potential in our countries. I believe that the modern role of high quality science in the solution of economic and social problems of a nation, including basic research, should not be entrusted exclusively to institutions of higher learning.

This tendency, common in many Latin American countries, must change.   The different branches of private and public sectors must, likewise, give serious consideration and stimulus to research, without establishing sterile and wasteful competition between local institutions. The ideal would be to create the necessary research facilities in specific fields in accordance with the actual historical necessities.

Thus, for example, if a country derives a great part of its national income from the exports of agricultural products, it should concentrate not merely on application of modern technology based on expensive equipment and materials to increase its production, but must accompany this technology with the most modern research methods, including the techniques of genetic engineering, based on the findings of physiologists, biochemists, and plant geneticists. Governments must create facilities for this purpose without hesitation, employing the available resources of established institutions within each country, or by instituting the necessary independent coordinated bases. In these cases, we must also take full advantage of international cooperation.

THE MAKING OF A SCIENTIST

Educational Problems

The first barrier we come up against is the generally poor quality of the teaching of science at all levels in Latin America. Our education is a fundamentally informative, commonly a passive and autocratic, method, the opposite environment of what the teaching of science requires. Instead of contributing toward the development of an open and creative neutrality – inquisitive, critical, and free from useless and complex prejudices – the system has led to the opposite results. At some times it has stimulated introverted personalities with limited aspirations, and at others, it has fostered undisciplined thinking, individual and collective irresponsibility, and the pursuit of material and superficial intellectual values. All these factors constitute open warfare on what should be the pursuit of an ideal and fertile scientific environment.

During the last few years, at least two important Latin American figures have described the failure of our educational system. Ivan Illich has condemned the conventional school system at all its levels, and accuses it of being the cause of a number of serious problems of our societies: it produces polarization and social stratification, contributes to the perpetuation of the status quo, serves to domesticate and alienate. Illich, in his demolishing criticism, further states that the school "limits the vitality of the majorities and minorities, emasculating the imagination and destroying spontaneity." Also, in emphasizing the authority of the teacher, it creates the image that one must learn from him alone, thus discouraging self-learning. He ends by definitely proposing the abolition of the conventional school as such.(4)

Paulo Freire's concerns coincide in part with Illich's, but on the oppressive structure of our elitist societies.(5) He is more positive in proposing solutions and methods that would sufficiently neutralize the basic errors of our systems of education which, in his opinion, dehumanize the components of our societies. Thus, Freire advocates an education capable of providing a critical and reflective attitude in the pupil, and claims that "alphabetization of man" should lead him from passivity toward the development of creative and inventive creativity, and therefore, originality and objectivity of the individual; but, if applied at an early stage, it would also permit the child to analyze rationally the world around him, adequately weight his/her reactions, and establish cause-effect relations that would aid him/her in facing everyday problems. This does not imply that all men and all women of a modern society should be scientists, but that the scientific method should be applied in all its scope to the problems of modern living, without discarding the inherent spiritual component of the human being. It should be applied to the formation of more objective individuals possessed of independent and ample criteria, capable of formulating hypotheses and of proving them, critical of themselves and of the society in which they live, and willing to accept changes and to approach problems with a reflective and creative mentality within the frame of social justice, human solidarity, and ecological balance. In true scientific education, the student will find a school of mental discipline that will guide him in the practice of seeking and respecting the truth, in acquiring a receptive attitude, and in banishing from his mind all kinds of prejudices.

The solution of such deep-rooted problems is not easy. Merely changing the formal scientific curricula will not likely have actual or positive and lasting effect. There must be fundamental changes in our systems of education, reorienting them toward the transmission of different attitudes, which cannot succeed unless they are accompanied by a corresponding change in the attitudes of teachers. I am convinced that this defect in our education is an important cause, and, at the same time, an effect of our scientific underdevelopment, and that it will be impossible to alter radically the attitude and behavior of our educational personnel without recruiting talents from abroad. An injection of new blood is required, not only in the field of modern teaching techniques, but also in the sciences that are taught. This would be a positive contribution toward new educational techniques, new approaches to educational problems, and toward a new orientation of modern scientific education that could disrupt the vicious circle in which we are presently moving.

Our only approach to this educational crisis would be through government programs oriented toward producing decisive, but basic and qualitative, changes in our educational system by means of heroic actions that would attack the very roots of the true causes of our educational failure. Only thus can we bring about positive changes in the way of thinking of our youth; and only thus can we correct the current statistical imbalance in many of our universities from being top heavy in nonscientific fields to a greater focus on scientific orientation of programs so that a higher percentage of our students will feel that their goals and vocations have been favorably supported.

It is only fair to recognize that in a few Latin American countries the first steps are being taken toward favorable solutions to the educational problems outlined above; unfortunately these are as yet insufficient. A positive approach would be the creation of centers for upgrading the sciences, such as those established in Japan in 1960.(6)

Such centers are useful in promoting the adoption and creation of methods and procedures for the functional teaching of the sciences, in keeping with the most advanced pedagogical techniques and with the collaboration of the most outstanding scientists of the community; the preparation of basic materials for active teaching institutions, operated without ostentation in the use of material resources, and not merely housing historical objects. They must transmit a humanizing, technical and scientific message easily understood by the majority of the visitors. Furthermore, a wide range of related extracurricular activities should be included for those students who show a marked preference for any of the branches of science.

The Scientific Career

As for the diverse aspects of university level training in the sciences, I feel that the educational institution should provide both the means of detecting and the programs for stimulating those individuals that show promise and ability for research. The highest priority should be assigned to programs that would further motivate these individuals at an early stage in their careers, and encourage them to continue post-graduate training in their respective fields. The above-mentioned training must undergo a systematic process of strengthening throughout universities in Latin America. Such a program should lead to the recruitment of "critical masses" of researchers in the various disciplines, thus avoiding unnecessary interinstitutional duplications within the same country. This can be achieved by means of an efficient university coordination system operating at either national or regional levels.

I believe that the preliminary formation of well-trained groups in the various disciplines should, in turn, serve as a "triggering mechanism" for the establishment of postgraduate programs at the Master of Science level which should, within a reasonable period of time, generate doctoral programs. A failure in this process would create awkward situations, such as the indefinite maintenance of programs which would eventually lapse into mediocrity under the pretext that they cannot demand a level of excellence akin to that of a doctoral degree.

Once a scientist has been trained, it is essential that there exist a dynamic environmental conducive to research and the pursuit of a career. This would permit a just and adequate stratification which should be concretely enforced thorugh economic and material stimuli. In this respect, the National Councils for Science and Technology, as direct representatives of their governments, must play a primary role in providing the necessary links between the various universities and institutions in achieving the objectives inherent in such programs.

In those disciplines in which a given country cannot develop high level research centers, primarily due to the lack of human resources, it becomes imperative that selected candidates should continue further graduate training abroad, especially in those fields of research with high priorities in the developmental objectives of the nation. The application of this type of program is of special concern during the initial phases in which young, outstanding candidates are called to complete their scientific training at either doctoral or other levels of specialization. This phase should preferably be of no less than three to four years. Far too often, shorter intervals of training cause more harm than good, particularly as regards the attitude with which such candidates return to their native countries.

Latin American nations should organize a well-planned scholarship program comprising national as well as international training alternatives. This would permit adequate rationale for stimulating the formation of the required human resources in the various scientific and technological fields. This program should be adequately administered and characterized by objective criteria based, in turn, on thorough analyses of existing priority needs within the overall context of the academic and scientific goals of the particular country, and preferably coordinated by each of the Latin American Science and Technology Councils with due consideration given to existing national plans of development. Furthermore, such programs should include adequate planning in order to take full advantage of the newly acquired knowledge of the returning scientists.

One aspect that has been disregarded in Latin American countries is the necessity of creating a special system for subsidizing postdoctoral work. This, however, is of prime importance in the formation of scientists endowed with knowledge and training sufficient to undertake creative research for themselves, as well as to participate in the direction and formation of future researchers.

In this context, we must keep in mind the words of Hans Krebs(7) in that the true manner of forming a solid scientist lies in offering promising candidates the opportunity to participate in high-level research centers for reasonable periods of time, under the direction of a true master researcher. It is precisely there that a creative and critical attitude is learned, together with habits of hard work, correct research techniques, and the courage to respond decisively to scientific problems with a strong devotion to the quest for truth. Such endeavours would expand greatly those fields of knowledge that have been deficient in their earlier level of development.

Precisely, the lack of a sufficiently favorable research environment has been a fundamental cause of the "brain-drain" phenomenon frequently observed in our countries. One can hardly be surprised when one considers the degree of isolation and incomprehension that scientists often have to face in their countries. It is obvious that measures such as those mentioned regarding contractual importation of foreign scientists must be simultaneously correlated with programs that provide an adequate environment of stimulation for repatriation of national talent.

Although statistics indicate that the brain-drain phenomenon constitutes a substantial loss to the resources of our various nations, I strongly feel that, without a reversal of this phenomenon, only a contingent of imported talent can adequately induce the required scientific development of Latin America.

Brazil offers an interesting example in this respect. At the beginning of the century, and prior to the creation of its universities, institutes such as the Oswaldo Cruz in Rio de Janeiro, Butantan, and the Biological and Agronomical Institutes of São Paulo were created through the visionary efforts of truly remarkable men. The names of individuals such as Oswaldo Cruz, Vital Brasil, Henrique da Rocha Lima, Arthur Neiva and others leap from the pages of history as creators of Brazilian experimental science.(8) The degree of success achieved by these institutes, under the guidance and inspiration of the master-researchers who created them, stands as an example of free, creative, and high-level research conducted by the most valuable human resources of the times.

Even at the beginning of the century, Oswaldo Cruz, the founder of the institute that bears his name, was able to stimulate the formation of the best scientists of his country whom he later sent to Europe and the United States to complete their training in their various specialties. Similarly, he was instrumental in opening the doors of his institute to a group of notable German scientists.(9) A parallel phenomenon occurred through the mediation of Vital Brasil and his Instituto Butantan, and was repeated in the Brazilian universities, particularly the University of São Paulo established in 1934. The latter undertook a deliberately planned importation of outstanding European scientists which, through their diverse fields, stimulated the true birth of the experimental and basic sciences in Brazilian universities.(10)

In recent times, the very same situation, though in the context of a more modern conceptualization, is being repeated through specific treaties between Brazil and the United States. These programs, primarily designed for joint research activities in the field of chemistry, utilize foreign talent in an organized and disciplined fashion through organizations such as the Brazilian CNPq and the United States National Academy of Sciences. The end result is intellectual cross-fertilization between outstanding young North American chemists and their Brazilian counterparts. These young researchers stay for several years in Brazil to work alongside native chemists in high level academic programs. Although this program was initiated in 1969, it has already borne important fruits. Through this effort, the field of chemistry in Brazil has advanced considerably, not only in its academic aspects but in its impact on modern Brazilian industry.(11)

THE PROBLEM OF RETAINING MANPOWER

Let us now mention some aspects of the productivity and preservation of Third World scientists. It is obvious that if scientists are indispensable elements in the structuring of development-oriented national programs, conditions should be created to guarantee their remaining within the confines of their native country.

The brain drain phenomenon has been thoroughly analyzed by a number of authors.(12) It represents one of the principal paradoxes in the development attempts in Latin American countries. Far too often, the individual researcher is unjustly blamed for seeking economically rewarding employment abroad.   Criticisms such as these tend to overlook the utter lack of stimulation, and sometimes outright persecution, with which native scientists must frequently cope. These problems become even more aggravating in those countries that suffer from chronic political instability.

If national authorities continue to overlook the necessity of providing adequately stimulating programs for native scientists, it will become increasingly difficult to assemble the essential manpower, whose absence constitutes the main bottleneck in the attempt to attain an authenic degree of development.

We must never forget that science constitutes a necessary ingredient in every modern society. On the one hand, it stimulates the formation of personnel capable of delineating creatively the principles of natural and social phenomena, thereby elevating the intellect to a superior plane and strengthening the cultural climate. On the other hand, it is the scientists who undertake the task of applying these known principles of reality in the service of progress, dignity, and social wellbeing. They constitute the shock troops that endeavor to overcome misery, ignorance, malnutrition, diseases, and all forms of social injustice.

In our countries the productivity of a researcher, in both basic and applied science, will depend upon the degree of esteem accorded the performance of efficient and useful scientific work, as well as upon the stimulation of the environment within which this work takes place.

CONCLUSION

One can hardly refer to technology as an isolated fundamental ingredient of present-day production, as if it could be entirely separated from its scientific component, particularly in the context of the twentieth century. Undoubtedly, it is a broad scientific outlook that constitutes the principal means through which a given country can generate new technologies as well as improve existing ones. Only such a transformation can create nations that have a competitive potential in the world market of knowledge. It is for this reason that I find myself concerned when I hear some intellectual espousing principles of technological development in the absence of a correlated scientific growth. It is essential to understand that no nation that does not seek to reinforce programs of high-level research can dominate modern technology or create indigenous know-how. Furthermore, basic research is necessary to generate and preserve indigenous scientific talent and induce a creative scientific outlook.

The political decision to incorporate science in a rational and consistent manner into the context of a particular nation has been prophetically expressed in the words of Chaim Weitzmann: "I feel confident that science will bring to this land both peace as well as a renewal of its youth, creating the fountains of a new spiritual and material life. In this respect I refer to both pure as well as applied science."(13)

In summary, it is upon human resources that our nations must place their principal efforts in the forthcoming years. This must be clearly and explicitly stated in the national plans for scientific and technological development which each country must elaborate.

NOTES

(1) M. Roche, "Factors Governing the Scientific and Technological Development of a Country." Sciencia 3 (1976): 75-84.

(2) M. Bunge, "Tres politicas de Desarrollo Cientifico y una Sola Eficaz." Interciencia 2 (1977): 76-80.

(3) A. Kalnay, "Algunas Observaciones Sobre la Planificacion de la Ciencia en Paises en vias de Desarrollo," Interciencia 2 (1977): 95-98.

(4) I. Illich, En America Latina Para que Sirve la Escuela? 3rd. ed., Ediciones Busqueda, Mexico, 1974.

(5) Paulo Freire.

(6) B. Glass, "The Japanese Science Education Centers." Science 154 (1966): 221-28.

(7) H.A. Krebs, "The Making of a Scientist," Nature 215 (1967): 1441-45.

(8) Henrique da Rocha Lima and M. Silva, "Birth and Development of Experimental Science in Brazil," Interciencia 1 (1976): 215-18.

(9) H.B. Aragao, "Noticia Historica Sobre a Fundacao do Instituto Oswaldo Cruz," Mem. Inst. Osw. Cruz 48 (1950) 1-50.

(10) Rocha and M. Silva, "Birth and Development of Experimental Science in Brazil."

(11) M. Frota-Moreira, B.K.W. Copeland, "International Cooperation in Science. Brazil-U.S. Chemistry Program," Interciencia 1 (1976): 139-46.

(12) I. Saavedra, "El Problema del Desarrollo Cientifico en Chile y en America Latina," Cuad. Real. Nac. 1 (1969): 32-52. H.M. Nussenzweig, "Migration of Scientists from Latin America," Science 165 (1969): 1328-32.

(13) Quotation on Weitzmann's tomb in Park of Weitzmann Institute in Rehovot, Israel.

II
Infrastructure and Technology Transfer

4 Preinvestment Work and Engineering as Links Between Supply and Demand of Knowledge*
Mario Kamenetzky

OLD AND NEW SEMANTICS

Technology

For capital, labor, and natural resources to be integrated into productive units to produce goods or services, it is necessary to know what to do with the resources and how to do it. The integrating factor is knowledge. A country with petroleum, for instance, can have capital and labor to develop it, but it can do nothing unless it also has the know-how required to extract it from the soil, transport it, and process it.

Scientific research creates knowledge about the resources and what to do with them. Technology deals with how to do it and may be defined as the organized set of all the empirical and scientific knowledge required for producing and distributing any good or service. Technological knowledge may be the result of either technological research or practice and tradition.

Engineering

Engineering services use scientific and technological knowledge to:

- design and build new production units; and
- optimize existing units and keep them in operation.

The fundamental economic activity of engineering is, therefore, the provision of services for production. It involves the transformation of

*I thank Brian Magee for his help in editing this paper. All errors are mine.

usable knowledge into used knowledge. In addition, engineers may also provide services for the creation of knowledge by taking part in scientific and technological research.

These two aspects of the engineer's professional work are explicit in the French language, in which the different engineering branches are identified with the expression Ingenieur du Genie . . . civil, mecanique, chimique, etc. Ingenieur derives from the old French word engine, meaning war machine. It is equivalent to the English engineer and to the Spanish ingeniero. It connotes skills in planning, implementing, and operating productive undertakings, that is dexterity in the use of knowledge for designing, building, and running engines and apparatus. Genie derives from the Latin genius and describes the creative aspects of the engineering profession, the skills required to perform scientific and technological research.

Consulting Services

The available knowledge often needs to be organized before it can be used. Traditionally, the organization of knowledge has been assigned to consulting services during the planning and preparation of investment projects.

The need to introduce a process for organizing knowledge previous to its use may have resulted from the exponential growth of accumulated knowledge. Statistical analysis of historical data shows that the body of knowledge has expanded very rapidly, doubling every 10 or 15 years.(1) In the United States there were 1,271 technologies in 1970 for a total of 748 processes in the manufacture of 545 chemical products. Thus, an average of almost two technologies was available for each type of process and for each type of product. However, there were certain processes for which up to 15 alternative technologies were available, and, excluding the 55 processes with only one possibility, the average number of technologies to be considered for each of the remaining was almost four.(2)

Consulting Engineering

We often find the term engineering modified by the adjective consulting. The adjective often introduces confusion, because it has two different connotations. By one of them, consulting refers to special services — precisely those defined above for the identification and preparation of investment projects. On the other hand, consulting may also connot a mode of providing any kind of technical services to production units or government bodies by independent agencies that gather and transfer knowledge by means of drawings; specifications; calculations; economic, financial, and social analyses; managerial or technical advice; etc.

Consulting engineering would, therefore, mean engineering services provided in a consulting mode and supplying knowledge derived from the engineering sciences for either the preparation and implementation of a project or for the production of goods and services.

A New Classification for Consulting and Engineering Services

To avoid confusion it is better to name the services required for the full development of projects and the operation of the resulting facilities according to their objectives.(3) Three main categories may be recognized (see Table 4.1):

1. pre-investment services for identifying and preparing projects. They require inputs from engineering and may be provided on a consulting basis.

2. project implementation services, with engineering and construction as their most relevant components.

3. services for operation and maintenance. They advise operators and administrators of production units on how to optimize the use of the embodied and disembodied knowledge transferred during the implementation of the projects.

WHY PREINVESTMENT SERVICES DESERVE SPECIAL ATTENTION IN LDCs

The quality of preinvestment work will have a large influence on the efficiency with which a country allocates its resources. This subject is of crucial importance to LDCs. Preinvestment work requires that the social and cultural environments of a project be adequately considered. Its main tools are methodologies for selection and evaluation of alternative technologies and products. Thus preinvestment work is chiefly probabilistic, cannot escape subjectivity, and demands the participation of local professionals and technicians.

The excellence of project implementation services, to a large extent, will determine the operational efficiency of the production units. Because their basic tools come from universal engineering sciences and mathematics, they are essentially objective and determinate and, hence, can be imported without affecting the suitability of projects to meet local conditions.

Furthermore, at early stages of development, when local engineering is not available for project implementation and operation, the local preparation of projects may reduce the risks of mistakes by providing the foreign engineering teams with adequate information on the local physical and economic conditions that may affect the design, and on the local social and cultural characteristics that may influence the operation of the facilities.

Fortunately, LDCs can more easily train their professionals and technicians to perform preinvestment work than project implementation

Table 4.1. Creation and Use of Knowledge

ACTIVITY	SERVICES REQUIRED BY THE ACTIVITY	RESULTING PRODUCTS
SCIENCE AND TECHNOLOGY	Scientific research	Scientific knowledge
Creates, adapts, stores, and distributes knowledge	Technological research	Technological knowledge
PREINVESTMENT WORK	Project preparation and evaluation	Market studies
Organizes knowledge	Preliminary engineering	Prefeasibility reports Feasibility studies
PROJECT IMPLE-MENTATION	Detailed engineering	Drawings, diagrams, specifications, models, calculations
Uses knowledge to implement projects	Procurement	Equipment and materials for the production facilities
	Construction and installation	Production facilities
PRODUCTION*	Training	Human capital of the enterprise
Uses knowledge to produce and distribute goods and services	Start-up operation	Goods and services
	Technical assistance	Improvements in production, organization and managerial technologies

*For production, only those services not provided by the permanent production team are listed.

work. A young professional can become acquainted more quickly with the methodologies involved in preinvestment work than with the detailed engineering of production units. For instance, the feasibility of a petrochemical project can be assessed by a team of generalists without any experience in that particular field, provided the team members are able to adequately sense the context of the project and obtain adequate information on technological alternatives and a sound preliminary engineering design. The detailed engineering design and the construction supervision of such petrochemical facilities normally require the inputs of people with experience in similar installations.

PREINVESTMENT WORK AND THE UNPACKAGING OF THE TECHNOLOGY

Any investment project encompasses a package of technological knowledge. Part of this knowledge is embodied in the capital goods required for the production process. Another portion appears in the form of drawings, specifications, maquettes, diagrams, and manuals. The remainder is directly delivered by the engineers and technicians who participate in the design and execution of the project.

Often a given technological package is selected at the prefeasibility stage of the preinvestment process. The subsequent feasibility analysis should disaggregate the initial package into its component unit operations, and for each one of them should select

- equipment and installations;
- control systems; and
- operation and maintenance practices.

Obviously, if a technology is to be specially created for a given project, preinvestment work (instead of the above described process of unpackaging) will build a package with existing pieces of technological knowledge plus those found while performing the technological research work required for the development of the new technology. In many cases, especially in the agricultural sector, even those projects which do not require a newly-created technology will need to build a new package tailoring existing knowledge to local conditions.

When relying on existing technologies, the investor, especially in the industrial and infrastructural sectors, may put them into use by following either of two fundamental modes:

1. The "turnkey" mode, in which the production installation is delivered to the investor already operating with trained personnel; and

2. The self-administration mode in which the investor himself, or in collaboration with a third party (such as an engineering firm), buys the different pieces of equipment and technological knowledge separately and then coordinates them.

In either case, it is convenient to disaggregate the technological package while performing the preinvestment work in order to adapt each operation and procedure to the local conditions. Also in either case we need to refer to the cost and availability of capital; the type, cost, and availability of natural resources; the skills, costs, and availability of labor; the possibilities of local supply for capital goods and engineering services; the psychological and cultural characteristics of the population; and the ecological balance.

When the investors manage the implementation of the project, they gain skills for preparing, designing, and implementing future projects.

In the progression from accepting turnkey operations to actually being the managers and prime contractors of this projects, the local investors may consider different levels of disaggregation of the technological package. At first – the most elementary – the investor asks for a list of each one of the supplies included in a turnkey operation and their prices. This would be a minimally assertive act on the investor's part to clearly and precisely know what he is purchasing.

In many cases, the learning process for the development of local consulting and engineering capabilities must start with this modest step towards a later, more complex, disaggregation.

At a more advanced stage, still under turnkey conditions, the investor requires that some of the goods and/or services in the list of supplies be purchased in the country where the project will be implemented, adapting their specifications to the possibilities offered by the local markets. The investor may do this in four ways:

1. letting the turnkey supplier use his own criterion when deciding where and how the local services and goods are going to be purchased;

2. providing him with a list of local suppliers who have the investor's confidence, and letting the contractor make the final selection;

3. specifically indicating where the goods must be purchased and with whom services are to be contracted; or

4. purchasing the local goods and contracting the local services on behalf of the turnkey supplier.

When the investors have developed certain skills and a feeling of self-reliance, they will be able to coordinate the complete project themselves. They may then –

– buy, license, copy, develop, or request the development of the necessary knowledge;

– design or request the design of the different operations involved in the selected production process;

- procure the equipment, materials, and services for executing the project where and how they may consider convenient; and

- train people to operate the installations.

MOTIVATIONS AND OBSTACLES IN UNPACKAGING

The full disaggregation of a technological package requires decision makers to assume substantial risks while their psychological motivations may lead them to prefer the much safer decision of discharging all the responsibilities on a turnkey contractor.

Compulsory regulations forcing decision makers to assume risks are ineffective. When individuals are forced to perform duties beyond their skills and training, they are likely to make mistakes in the design or the implementation of the project. This puts the individuals under unusual stress, and leads them to find ways of reducing the stress through bypassing the regulations. For instance, when imports of detailed engineering are forbidden, the prices for equipment are often artificially increased so as to cover the cost of the engineering of the corresponding installation which still comes from abroad.

Education is better than repression. Professionals and entrepreneurs should be trained to recognize risks, to accept them as an integral part of the investment project, and to minimize them through efficient preinvestment work and project organization.

For this purpose, individual and social behavioral patterns that may oppose a risk-taking attitude should be identified. Some of the individual motivations leading investors and/or their staff to prefer turnkey packaged operations are the following:

- The desire to avoid any new external factors that may disrupt one's intrinsically fragile, psychological equilibrium. It is easier to cope with difficult situations that may arise during implementation when the main responsibility of a project lies on an external, experienced firm than when it is assumed by oneself.

- Self-repression of originality by fear of "burning one's fingers."

- Hopelessness regarding oneself and high hopes placed on the others.

- Fear of facing scarcity situations because of losing one's job, for instance.

- The desire to accumulate wealth without effort: tasteful bribes are more easily obtainable when negotiating a whole project with a few foreign contractors, than when dealing with a multiplicity of local suppliers.

Socioeconomic and cultural factors also opposing the unpackaging of technologies are, among others, the following:

- Higher costs and uncertainty in the supply of local services, parts, and equipment.

- Lack of skilled local consulting and engineering teams for performing preinvestment work.

- Inadequate formation of professionals and technicians by universities which are often disconnected from local reality and submit to destructive political pressures.

- Lack of awareness of the multiplier effects that the disaggregation of a technological package and its adaptation to local conditions may have on a national production system.

- Frequent requirement from external aid channels to buy capital goods and project implementation services from developed countries. (The eruption in the financial markets of low-developed money-lending countries may paradoxically reinforce this tendency, because part of their capital surplus is invested in developed countries in world renowned equipment manufacture or project design firms.)

- The lack of support from many technology suppliers in helping the buyers to acquire the mastery of the knowledge by directly involving them in the design and implementation of the project.

On the other hand, certain individual motivations and socioeconomic factors may also favor the full disaggregation of the technological packages and their adaptation to local conditions. Among them, the following may be found:

- The psychological rewards professionals, technicians, and entrepreneurs experience from their involvement in the design and implementation of a project.

- The memories accumulated in developing societies of difficulties suffered because of the use of imported technologies that were not adapted to the local conditions.

- The spreading of national strategies promoting the local supply of capital goods and engineering services.

- The growing number of competitive technologies, and of their sources of supply, facilitating the search for sellers or licensors willing to help the buyer or licensee in the unpackaging of the technology.

- The increasing encouragement by international development organizations that local teams perform the preinvestment work in projects they intend to finance.(4)

Conscious and unconscious resistances to take the risks involved in unpackaging are great. When decision makers are able to overcome their internal blocks, they may find the ways to neutralize the adverse socioeconomic and cultural factors.(5)

UNPACKAGING AND THE LOCAL PROVISION OF CAPITAL GOODS

The substitution of local capital goods for imported ones is crucial for many developing countries due to their lack of foreign currencies, the continuous deterioration of the foreign trade terms, and their need for equipment and tools adequate to their factor endowment and market size.

In Argentina, for instance, capital goods represented approximately one-third of the total imports in 1972. In a 13-year period (1960 to 1972), the amount spent on imported capital goods totaled $4,560 million, the yearly expenditures varying from a minimum of $154 million (1965) to a maximum of $556 million (1962).(6)

In Mexico, in 1971, Petroleos Mexicanos was still importing 40 percent of the process columns, 55 percent of the pressure containers, 70 percent of the heat exchangers, 14 percent of the pumps, and 70 percent of the compressors required to implement its petroleum and petrochemical programs.(7)

In any country, the increase and diversification in the local supply of capital goods parallels the strengthening of the national preinvestment and engineering teams. As indicated by Perrin,(8) the project engineering is always in close relationship with the equipment suppliers, and the latter's design and production capacity conditions the degree of development of the former.

This shows, once more, the importance of building local preinvestment services at a very early stage of a country's development. When there is not yet a local supply of capital goods and project engineering services are still undeveloped, the preinvestment studies should at least state preferences for certain foreign equipment, aiming among other things to standardize them as much as possible.

Standardization is always important, because it reduces the capital demand for stocks of spare parts and stand-by equipment. Its consequences are even more significant when the equipment is imported. Standardization reduces the consumption of the often scarce foreign currencies and prepares the way for the local manufacture of the same kind of equipment in the future. The local supply of capital goods and their parts, difficult in itself in small markets, is made still more complicated if the demand is fragmented by the use of different types of equipment for the same purpose. This reminds me of one case in point where three different processes, turnkey contracted with three different foreign suppliers and aimed to be integrated into a single petrochemical complex, were equipped with three different types of control instruments and electrical devices which accomplished identical functions.

At more advanced stages of development, local engineering and capital goods industries may begin to design and build the necessary equipment for those ancillary services common to many different industrial, agricultural, mining, and infrastructural projects – for instance, equipment for steam generation, fluid flow networks, storage facilities, and water and sewage treatment.

At even more advanced stages, local suppliers may also produce equipment for those operations that repeatedly appear within the different core technologies themselves – for instance, equipment for phase separation, heat exchanging, welding, or metal working.

In these last two stages unpackaging of imported technologies is mandatory if local production is to be promoted. The preliminary engineering of the project should determine those operations in the acquired technological package for which local equipment is available and should be used.

In large projects, if the potential demand for local goods and services were publicized immediately after the preliminary engineering is accomplished, the local suppliers would be able – while the feasibility studies were being conducted and the implementation decisions taken – to plan a response to the new or increased demand. Thus, the construction of new types of equipment or the production of new materials might be encouraged.

For instance, the Argentine affiliate of a transnational corporation once publicized its capital goods and installation requirements for a five-year period. The main beneficiaries of this announcement were small workshops in the area where the plant is located, which expanded and adapted their facilities in a joint move towards fulfilling those requirements. Also, large Argentine manufacturers of heavy equipment made good use of this anticipated procurement program.

PROCUREMENT OF ENGINEERING AND TECHNOLOGY

As has been previously described, technology can either be specially developed for a project or chosen from the stock of available packages. The engineering of the project can be performed by the provider of the technological package, by foreign or local engineering firms, or by the sponsor of the project.

Many enterprises, especially those producing final consumer goods, prefer to rent existing technologies through payments proportionate to their sales or, what is equivalent, to the number of times they have used the rented knowledge.

Others, mainly those manufacturing basic and intermediate products, usually pay the rent in advance, on the basis of a previously agreed maximum production rate, regardless of the period of time during which that rate or a lower one will be maintained. But, whenever the maximum rate is surpassed, the anticipated rent must be proportionately increased.

There are very few cases in international technology transfer in which knowledge is actually sold, giving a buyer full property of the technology and, hence, the right to use it at his will and convenience. Somewhat more numerous are the cases in which the temporary possession of knowledge, through the payment of royalties or lump-sums, allows a licensee to acquire the property of the improvements he may introduce into the rented knowledge while using it.

The combination of the various modes by which technology for a project may be acquired and the diverse ways in which its engineering may be executed are shown in Table 4.2. Each combination has different benefits and risks associated with it.

Figure 4.1 qualitatively shows the level of potential social benefits and of entrepreneurial risks associated with each engineering-technology-procurement combination. Positive effects on the dynamics of the development process increase when at least one of the terms in the above combinations is local. The positive influence of local components decreases when the services are provided by either research or engineering teams captive within the enterprise. In these instances, the learning-by-doing process cannot spread to the remainder of the productive system as easily as would be the case if the work were performed by local independent groups. The levels assigned to each combination only represent the author's qualitative judgments and reflect his personal experiences with many of the described combinations.

CONCLUSIONS

Consulting and engineering services establish the linkage between the supply of knowledge and the demand for its use. It is proposed here that consulting services be correlated with preinvestment work, while leaving the term engineering to identify the design and calculation work required to implement a project (detailed engineering). Engineering work is also needed during the preparation of the project for evaluating technological alternatives at the prefeasibility stage and for performing the preliminary engineering at the feasibility stage.

Preinvestment work defines the technological choice and, hence, the allocation of resources. For this reason, the development of groups performing preinvestment work should have the highest priority in the process of building local technological capabilities.

For certain projects, and as part of preinvestment work, it may be necessary to build technological packages by using existing knowledge as well as knowledge specifically created through technological research. For other cases, the most convenient option is to acquire existing technological packages. Preinvestment work should then disaggregate the packages into their operational components in order to adapt each piece of hardware and software to local conditions and in order to determine the materials, equipment, and services that may be procured locally.

As stated above, preinvestment work requires inputs from engineering in order to build or disaggregate technological packages. It also demands inputs from economics and the social and natural sciences for sensing the physical, economic, social, and cultural conditions that will influence the technological choice.

Either when building a new technological package or when performing the disaggregated analysis of an existing one, the contribution of local preinvestment work to self-reliance and the efficient allocation of resources is reinforced by the increasing participation of local research and engineering teams.

Table 4.2. Engineering and Know-How Procurement Combinations

Engineering executed by ⟶ Technological knowledge acquired through	the seller or licensor of the technology	foreign engineering teams contracted by project sponsor	local engineering teams contracted by project sponsor	engineering teams working inside the enterprise sponsoring the projects
renting with payments proportionate to sales	1*	2	3	4
renting with advanced payments as per maximum production rates	5	6	7	8
purchasing	9	10	11	12
developing a special technology for the project through a contract with a foreign research team		13	14	15
developing a special technology for the project through a contract with a local research team		16	17	18
developing a special technology for the project by a research team working inside the enterprise sponsoring the project		19	20	21

* Combinations have been numbered so as to facilitate the comprehension of Fig. 4.1.

Unfortunately, the risks for the entrepreneurs increase when, instead of using an already-proven, packaged technology, they share the uncertainty of developing a new technology and of training local technological capabilities.

In order to make readily available to society the positive effects of unpackaging existing technologies and of creating new ones, the governments should back the development of local consulting and engineering services; support entrepreneurs in facing the risks; reward the use of technologies locally developed, or locally adapted; and create large markets for local services. For these purposes, the public sector may use its purchasing power and appropriate fiscal and financial incentives.

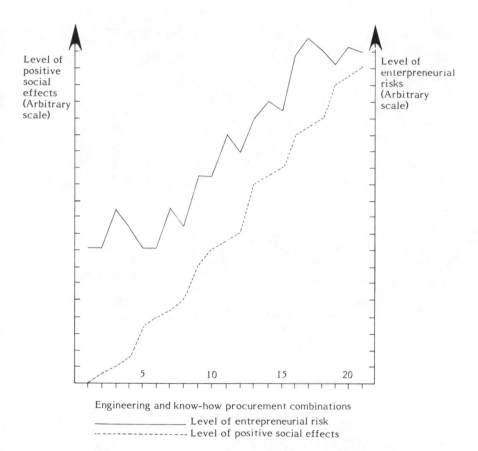

Engineering and know-how procurement combinations

——————————— Level of entrepreneurial risk

----------------- Level of positive social effects

Fig. 4.1. Potential Social Benefits and Entrepreneurial Risks
Associated with Different Engineering/Technology
Procurement Combinations

Local preinvestment work and engineering have proved to be an efficient linkage mechanism between the supply of foreign, large scale, modern technology and its local demand. They help local enterprises in mastering the imported knowledge. But consulting and engineering services, at least in their traditional form, may not be the best channels for the transfer of small scale, poverty-oriented, ecologically-minded, participative technologies. Perhaps nothing can replace a direct exchange at grass-root levels – a cooperative transfer of skills and knowledge from people to people. This is an issue worthy of study and experimentation.

NOTES

(1) Derek de Solla Price, The Relations Between Science and Technology and their Implication for Policy Formation. (Stockholm, FAO: 1972), p. 9.

(2) "Process Technology for License or Sale," Chemical Engineering, (April 20, 1970), pp. 114-144.

(3) This was the criterion adopted in a team meeting of the Science and Technology Policy Instruments Project, sponsored by the International Development Research Centre of Canada, held in Caracas, Venezuela, November 1975.

(4) See: Uses of Consultants by the World Bank and its Borrowers, April 1974, Item 1.16; Uses of Consulting Firms by the Interamerican Development Bank and its Borrowers, 1975, Section 2.08.

(5) Analysis of some cases where this really happened in Latin America may be found in: Marcelo Diamand. Las Posibilidades de una Tecnología Nacional en Latinoamérica, Latin American Forum on Technological Development, Austin, Feb. 1975; Marcelo Roberts, La Innovación Tecnológica en Latinoamérica, OAS, Washington, 1972; Jorge A. Sábato and Oscar Wortman, Apertura del Paquette Tecnológico para la Central Nuclear de Atucha (Argentina), OAS, Washington, January 1974, (mimeographed); and Mario Kamenetzky, Ciencia y Tecnología Argentinas en la Industria, Fundacion Bariloche, Buenos Aires, April 1972 (mimeographed).

(6) OECEI, Argentina Economica y Social, Buenos Aires, 1973, Vol. 2, p. 449.

(7) Lic. Arturo del Castillo, lecture on "Equipment, Demand for PEMEX's Plants," Memoria del I Congreso de la Asociacion Nacional de Firmas de Ingenieria, Mexico, 1971, p. 132.

(8) J. Perrin, Design Engineering and the Mastery of Knowledge for the Accumulation of Capital in Developing Countries, IREP, Grenoble, 1971, p. 8, (mimeographed).

5 Multinational Corporations and the Transfer of Technology
Mira Wilkins

Less developed countries have looked to multinational corporations (MNCs) for technology. Most less developed countries recognized – implicitly, if not explicitly – that there is nothing automatic about technological change. Someone or some institution must introduce a technology (or technologies), apply the process, and produce the product. Somewhere there must be a market for the goods and services produced to sustain the use of an introduced technology. A technology can be introduced in a physical sense (i.e., there can be present and available technological know-how or processes), but if there is no market for the goods produced in a locale at those costs of production then the technology will not be used at all (or will be abandoned if introduced).

A multinational corporation is one institution – one vehicle – for the transfer of technology across borders. It is by no means the only vehicle. Others include individuals (acting individually, or connected with universities, governments, or host-country companies), equipment manufacturers, universities, governments, and, of course, written words (journals, patents, or blueprints). Moreover, within the framework of a multinational corporation are individuals, who participate in the transmission process, and blueprints and patents that describe processes. The multinational corporation – as a vehicle – can also be associated with transfers by equipment makers, as well as by university and government personnel. For example, MNCs have close ties with equipment suppliers, hire university-trained personnel, and do research on government contracts.

A multinational corporation can transfer technology in a variety of ways from direct foreign investment with 100 percent ownership, to joint ventures, to no ownership at all, and licensing and/or management contracts. MNCs can also transfer technology merely through the export of products. Through the demonstration effect, products can (if the technology is not complex) be copied in the recipient country. The technology transferred can be its own technology or an acquired

57

technology; it can be a part of a process or the whole process.(1)

There is a literature on why a multinational corporation chooses to invest rather than continue to export.(2) There is also a literature on why a multinational chooses to have 100 percent ownership versus joint ventures, versus licensing or management contract arrangements.(3) Why a multinational corporation selects a particular route is not our concern here. We are going to take as a given the existence of these alternative forms.

What is distinctive about a MNC as a vehicle of transfer is that it has normally worked with a given technology; it does not simply have patents or blueprints. It has experience with the technology in practice and not in only a single facet (as is the case with most individuals with modern technology), but in all its facets. It has gone through trials and errors. It has purchased or built the plant and equipment used; it has hired and organized labor to use the technology or technologies; it knows the level of skill required; it has learned the difficulties with the methods, and presumably acted to eliminate them; it has typically developed quality controls on the technology; and it has produced and sold the product. It has, or can easily obtain, the know-how, the skills, and the managerial ability to translate the process or make the product a reality. The MNC knows about and has "logistical support systems, such as maintenance services and spare parts availability."(4) It is not simply "proprietary" technology, or secrets, or patents that it "owns" or "has"; it has not only worked with one aspect of the technology; but most important it has experience with all facets of a given technology. This advantage, we would suggest, is unique to business enterprises and not typical of other vehicles of technology transfer.

In this chapter I want to look at some of the public policy choices that less developed countries have in their search for technology, specifically in their relations to multinational corporations. I will look at the controversies over MNCs as agents of technological change.

But first, a few more words on the multinational corporation. There is no accepted theory of the MNC. We must recognize it as a business enterprise, with foreign direct investments, seeking markets and sources of supply. For our purposes, we do not have to decide whether its role is that of maximizing profits, or sales, or growth, or simply satisficing. We must, by definition, accept that it is in business to make a profit, and sales result in profits.(5) We must also accept that its cost structure is different from a national firm. Through its internalization of markets (by vertical and horizontal integration and diversification across borders), its information and its transaction costs should be substantially lower than those of a national firm.

By definition, a MNC is usually larger than a domestic enterprise. Most multinational corporations have first done domestically what they now do internationally. Most have accumulated capital, personnel, skills, technology, marketing organizations, and managerial expertise. They can devote substantial money to research and development and still have that activity represent a small portion of their budgeted activity.

Often the literature on technological transfer is vague on the issue of transfer and diffusion. From the less developed country's standpoint, "effective" technological transfer (or diffusion) involves not simply the physical introduction across borders of a new technology, but the absorption or adaptation of that technology to national needs.(6) Keeping this in mind, we come to the basic issue of our paper: Is the multinational corporation an appropriate vehicle for the transfer of technology to less developed countries? Perhaps we will find that it is the appropriate vehicle for transfer, but not for diffusion. Remember, however, diffusion of existing technology – effective transfer – is impossible without simple transfer. Thus we are back to the query, now modified. Is the multinational corporation appropriate in any facet of the transfer process? The response becomes a cacophony of "yes," "no," and "maybe."

Those who argue "yes" take the view that 1) the most modern technology is in the developed world; 2) in industrial nations, multinational corporations are in the forefront in owning and developing (if not in inventing) the leading technologies; 3) indeed, it is generally accepted that one, and for some economists the key, advantage that multinational corporations have is their technological edge;(7) 4) less developed countries would be shortchanged if they did not adopt the most modern technology; (8) therefore 5) if less developed countries do not receive the technology from the most obvious repository of such technology, they are not taking advantage of the best existing vehicle for transfer. The argument, of course, assumes that an owner (holder) of technology has lower costs of transfer than an intermediary and the elimination of a "middleman" is cost-reducing. It also assumes a technological neutrality: that social and economic differences in less developed countries do not figure in the equation.(9)

By contrast, the "no" argument concedes that 1) the most modern technology is in the developed world, but argues that what is currently the "most modern" may be inappropriate for less developed countries; 2) multinational corporations have technological leadership but based on capital- rather than labor-intensive methods which are more appropriate for less developed countries; and 3) the technological advantage may be present for MNCs, but rather than being an aid in development, they suppress indigenous technological achievement. 4) The "no" view does not accept that less developed countries would be shortchanged if they don't adopt the most modern technology; quite the contrary, it is appropriate, rather than modern, technology that LDCs should seek. 5) It accepts the premise that there are a wide variety of sources of technology; LDCs should not turn to MNCs, which, in their quest for profits, are bound to act in their own interests at the expense of the less developed countries. The argument, of course, assumes that the LDC has options in the choice of good vehicles for the obtaining and transferring of technology, and that at least one of these options has less cost than the use of the MNC as an agent for transfer. It also assumes that the problems of development in LDCs "differ basically" from those experienced by industrial nations.(10)

The "maybe" argument waffles on all these issues. At times, it is not a single argument but a continuum between the "yes" and "no" poles. It refuses to be boxed into simple answers to complex questions and looks to nuances. It accepts that the most modern technology is in the developed world. However, it argues that a less developed country must select <u>some</u>, but not all of that technology. The choice can be by industry, by process, or even by parts of a process. The choice must be based on the specific technology and the developing nation's economic plans. Others would rely on the market mechanism for the selection, recognizing, of course, an imperfect market. In each case, however, what is appropriate emerges from local conditions. Some processes may be transferred verbatim; others have to be modified. The "maybe" group differs from the "yes" group, which sees industrial countries as the exclusive source for technology, and from the "no" group, which is convinced of the inappropriateness of most, if not all, of modern technology to less developed countries.

Again, the "maybe" group would agree that multinational corporations have, or have access to, the most advanced technology: but since it is arguing for a selection of technologies, it has to move to the basis of choice. It grants that MNCs have, in general, technological leadership based on capital-intensive methods, but it is wary of too dogmatic an assertion that labor-intensive methods are always most appropriate to less developed countries. It points out that industries differ widely, and also that labor in LDCs lacks skills and, for that reason, labor-intensive processes may have hidden costs.(11) The "maybe" group notes that, at any point in time, for some processes there are many alternative technologies (with varying degrees of labor intensity), whereas there may be no alternatives for other processes.(12) Rigid rejection of MNCs because they may not use existing labor-intensive technology, or because they do not innovate in creating new labor-intensive technology, is shortsighted.

Granted that MNCs have a technological edge, but technologies develop within contexts of particular markets and environments, according to the "maybe" group. Appropriate technology that gives this advantage in one country may not provide similar advantage in another country. It must not be assumed that because multinational corporations have a technological advantage that this advantage is a universal, global, enduring edge, or that no other technology will suffice. A technology suitable to supply a high-income market may be unsuitable to the demands in LDCs. A technology developed in a country with one set of resources may be inappropriate in a different environment. In defining "local conditions" it is important to look at the market for the output (is it to be solely domestic or is an export market envisaged?); what are the local resources that are available for use? Dr. Wionczek writes of the "reformist" school that believes

> technology imports via MNCs are (useful as long as they are) directed and controlled by the host countries in such a way that they contribute to the creation of domestic scientific and technological capability . . . Host countries should require MNCs to make a reasonable contribution towards product and process innovation, of the kind most suited to national or regional needs . . .(13)

Dr. Wionczek is suggesting that not only should the multinational corporation scan existing technologies for the ones most appropriate, but that it should contribute to technological innovation physically within the host nation – under the direction of the host countries. Here, perhaps, we are getting to one of the most confusing parts of the argument. Here there is the least communication between the "yes" and "no" poles and the most fertile ground for the "maybe" position. The unequivocal advocates of MNCs as an instrument of transfer assume that the physical place that technological innovation should occur is where it is least costly, i.e. where the process is first initiated, where there are engineering and (for some industries) scientific skills, where the demand first appears. Technological innovations can then be shared. If adaptation is profitable it will be made. The "no" argument protests that multinational corporations are thus suppressing indigenous scientific and technological "potential." Dr. Wionczek would cope with this by pushing multinational corporations to contribute to domestic scientific and technological capabilities. The words "potential" and "capability" are key. Capability means the introduction of technology and the development of potential. Potential means the ability to innovate.

Some, including the present author, believe it is for the local firms to provide the basis for locally-applicable innovations. The MNC is most effective as a conveyor and modifier (on a voluntary basis) of existing technology.(14) Moreover, if the multinational corporations' technology is inappropriate in a less developed country, local enterprise with "appropriate" technology will have a competitive advantage. Rather than being suppressed they will flourish. Some of the "maybe" group – this author included – are doubtful that the presence of MNCs puts down indigenous technology. There is clear evidence that in countries where MNCs are present there has been an upgrading of local skills and talents, a training of local suppliers in new technology, and an emergence of modern marketing methods (in this case an introduction of social technology).

In some industries, the presence of multinational corporations may not stimulate indigenous technological development in the less developed country in the specific industry of the giant enterprise, but may spur technological development in related industries. Demand provides the basis for the choice of appropriate technology. To obtain quality inputs, the MNC provides technological assistance to local suppliers. It seeks low cost, quality supplies; its interests and its suppliers are joined in the market place.(15) To secure sales, it teaches dealers distribution and servicing methods.

Some commentators point to the growth of these "associated" activities as offering a basis for effective diffusion of existing technologies. They see the multinational corporation as providing for the creation of new national industries that serve the company and, in addition, aid economic development.(16) Critics view the growth of such activities as "satellization,"(17) believing the linkage effects result in permanent dependence on the MNC. Proponents believe such indigenous industries will have continued lives of their own, separate from the foreign investor. They see the suppliers and dealers as providing basic, needed infrastructure within the LDCs.

The "maybe" group accepts that "appropriate" technology is desirable in less developed nations, but it is not convinced that the multinational corporation always introduces "inappropriate" technology. Who decides what technology is appropriate? How and by whom does modification occur? Can modification, or better still extensive modification, be done most efficiently by the multinational corporation?

Some argue that as a matter of public policy the LDC government must "force," "coerce," "frame public policy in such a way" as to make multinational corporations introduce "appropriate" technology. Others argue that if it is profitable to modify technology, the MNC will do so without being coerced. My colleague, Dr. Thomas, has suggested that because the goals of the LDCs and the MNCs are different, what may be appropriate for the LDCs and most profitable for the MNCs will be different; and, unless "forced," the MNC will not adopt the most appropriate technology.(18) But here we are faced with the question of who determines "appropriateness"? Is a host government's decision on appropriateness necessarily more intelligent than the decision of a business that has experiences with various technologies? The evidence for this proposition is far from available, and contrary evidence does exist.(19)

It would seem that the MNC has its advantage based on its and others known technology. Modification to meet local conditions involves cost. If LDCs — governments or businesses — desire modification, if they desire to develop technological potential and "appropriate" technology, if they believe they are able to judge "appropriateness," then local, national entities within the LDCs should proceed on that basis.

In this view, it would be inefficient to expect (or to attempt to force) the MNC to invest extensively in redesign or to any great extent in LDC research and development. This point of view would maintain that the MNC might be a highly effective vehicle for the transfer of existing technology, but that "forcing" it to adapt would be inefficient in terms of national policy. It would also suggest that if domestic technological capacity is to develop — i.e. capacity not merely to imitate but to innovate — then the burden must lie not on the multinational corporation but on less developed country units (public or private).

The MNC is an easy target for blame for all the woes of the developing world. A public policy that insists that MNCs adopt some "appropriate" technology indicated by a governmental body, or that they establish research and development units within LDCs to ferret out the most "appropriate" technologies seems to this author a shifting of responsibility and a surefire door to permanent dependence by less developed countries. If a country desires to develop appropriate technology, it must learn from the available worldwide technology (and here the multinational corporation is an effective vehicle), and then make its own modifications.

Less developed countries, some of the "maybes" assert, will not by definition be shortchanged if they do not adopt the most modern technology, but they will be shortchanged if they do not adapt the most up-to-date technology.   They ought to be aware of the modern technology and be able to justify, in terms of long-range benefits, their adoptions and adaptations. Alternate technologies should be evaluated in terms of industrialization goals, costs (including information, research and development, and transaction costs), and market plans. The multinational corporation has, at present, greater access to information about alternative technologies in its specific industry than a local company or the host government. This is a resource that should not be neglected.

Multinational corporations are, indeed, the most obvious source of modern technology. They have major advantages as repositories of specific technology.   MNCs try out their technology first in developed countries. The "try out" costs include research and development, design costs on new machinery, search costs for appropriate suppliers, market studies, and so forth. Many of these are sunk costs. They do not need to be respent (at least to the same extent) when technology is transferred without change to a less developed country. To alter a technology may involve substantial cost. This varies by industry and by process, or part of a process, within an industry. The cost of modification may well be higher for the MNC than for nationals in the recipient country. Perhaps it is the task of the MNC to transfer known technology, and the task of the recipient nation or national company to take the initiative and adapt – in short, to accomplish the effective transfer, that is the diffusion. Perhaps the classic case where this has occurred was in Japan. There, in the postwar period, national companies bought technology from foreign multinational corporations and then modified it – introducing appropriate changes.

As noted, the multinational corporation is not the only vehicle to introduce foreign technology into less developed countries, or to innovate. LDCs have choices. The problem is that other vehicles of transfer may "cost" less in the short-run, but in the long-run may fail to be as effective as the MNCs.   For example, LDCs have often looked to "technical experts" from developed countries sent by developed country governments, by the United Nations (or one of its specialized agencies), by universities, or obtained through the job-market. Such individuals may be ex-employees of multinational corporations. They vary in knowledge of any particular technology. They usually lack familiarity with all the facets of a particular technology that is, by definition, present in a multinational corporation

Vehicles for transfer include machinery suppliers. Firms in the developed world are their largest buyers. Their designs are often linked to the needs of their most important customers rather than to particular requirements in less developed countries.(20)

Another vehicle for transfer is the written word – the blueprint, the academic or technical journal, the description in the patent application. The recipient of this information in the LDC must then translate the written work into a viable technology. This is difficult to do. Another agent of transfer is an individual from the LDC who goes abroad on behalf of governments or companies or universities in search of technology. Again, there is the difficulty of individuals' tranferring an advanced technology.

Technology transfer is never without cost. The costs can, however, be difficult to ascertain. How does one measure the "cost" of a multinational corporation's transferring its technology to a less developed country? What costs are relevant? What costs are reasonable: The total cost of developing the technology and transferring it? Of course not. A portion of the cost of developing it and all the transfer costs? But how does one decide what portion? The marginal cost of transferring a known technology? Perhaps. The profits on exports lost, if the technology is transferred and production begins abroad? Maybe, but over what time-frame?

It has been suggested that the way to evaluate costs is by "unbundling," by a recognition that similar technologies should be similarly priced.(21) If the LDC is in a position to obtain other bids, then there is no problem here, and if the technologies are in fact "similar." On the other hand, we would suggest that to price according to "disembodied" technology is unrealistic. The advantage that the MNC has as a vehicle for transfer is in its combination of technology with other elements in the package. A technology licensed by a MNC is not the same as the technology introduced by a handful of technical experts.

Licenses carry technology transfer prices. How licensing fees are determined varies. Do they reflect costs? What costs? Rather they seem to reflect, in the case of MNC's headquarters and subsidiaries, negotiated values (transfer prices within the MNC), and, in the case of multinational corporations and independents, market values. A multinational corporation will not sell to an independent if the technology is new (proprietary) and the price is not high enough. Prices clearly differ from cost and reflect both supply and demand factors. Price is what the MNC or other transfer agent will sell for and the recipient will pay. The "maybe" group assumes that the process of development in LDCs does not differ so fundamentally from that in developed countries, and therefore believes that the MNC has much to offer.

As noted earlier, a multinational corporation can transfer technology through direct foreign investment with 100 percent interest. For the less developed country, this has certain advantages: 1) the multinational corporation takes full charge of the activity; 2) the multinational corporation pushes for the highest profits (which can, of course, be taxed and provide tax revenue to the host country); 3) the multinational corporation (aware that the technology is within the organization) is less hesitant about the transfer to the less developed country; 4) the costs of transfer are internalized; the technology is, of course, not a free good; the payment is in the profits to the multinational corporation; and 5) markets for the output seem more secure than with other transfer forms.

A joint venture can involve control by the MNC, 50-50, or control within the LDC. Many LDCs, have been attracted to joint ventures because 1) they want their nationals to share in running the business; 2) they want their nationals to share in the profits (through means more direct than taxation); 3) they want their nationals to encourage more technology transfer; 4) they want their nationals to press for more local research and development; 5) they want their nationals to check on transfer prices; and 6) they feel with joint ventures multinationals are less apt to divide markets to their disadvantage.   Although joint ventures have appeal to LDCs, there seems to be little evidence to indicate that they result in better, more effective technological transfer.   It is true that joint ventures between multinational corporations and companies in Japan did serve as mechanisms of very effective technological transfer and diffusion in the post-World War II period, but these joint ventures were accompanied by an addition that LDCs rarely see as accompanying their joint ventures: the Japanese were prepared to pay a very high price for technology in order to obtain the best. By contrast, much of the efforts of LDCs has been designed to reduce the price of technology.

A multinational corporation can transfer technology to a less developed country through management contracts, service contracts, and licensing. Here the LDC pays a fee. The management contract is far from a panacea.   The experiences of India with the Managing Agency in earlier years make one hardly sanguine about this form.(22)

I am not suggesting that LDCs ought to rely on 100 percent ownership by MNCs. Clearly, in today's world that is a political impossibility. On the other hand, it is wishful thinking on the part of a less developed country to assume that because a joint venture exists, or because there is local ownership, the problems of appropriate technology are, by necessity, solved.

Indeed, in evaluating the MNC as a vehicle of technological transfer, the questions must always be asked, irrespective of the ownership issue, what are the advantages to the less developed country and what are the advantages to the multinational corporation? The MNC is a profit-making unit. The LDC has to select means of using the technological offerings of the MNCs in ways that provide a mutuality of rewards. For the LDC, it is possible that the forms that do not involve direct foreign investment (licensing, management contracts) can turn out to be more, rather than less, costly. Even so, these forms − that involve relations with multinational corporations − may be better bargains than the use of other, less appropriate, vehicles of transfer.

Perhaps our discussion is too general. Each LDC has distinctive technological requirements. Perhaps there is no basis for generalization. We do not allow this; we argue that there are − have to be − means of determining requirements. We recognize that the size of the markets in El Salvador and Mexico is totally different; the natural resource bases of Trinidad and Barbados appear to be unique; the capital resources of Venezuela and Colombia are far apart; the industrial structures of Aruba and Haiti are dissimilar; and the labor force skills of Costa Rica and Guatemala are diverse. An "appropriate" technology for Jamica may be inappropriate for Honduras.

Accordingly, what each country may ask of, expect, or obtain from multinational corporations will vary. Yet each will recognize that the multinational corporation must be in a position to make a profit or it will not invest (or make contracts). Each country must realize that no matter how technology is transferred, it will not be a free good; it will have a price. Each country should recognize that the MNC has major advantages over other vehicles of transfer (since it has experience with the "package" of management, skills, technology, marketing, etc.).

Effective technological transfer (diffusion) comes from a combination of imitation and innovation.(23) Perhaps LDCs should use MNCs to give them the basis for imitation. They should expect nationals of their own countries to innovate – developing novel technology and shaping the foreign technology to national needs.

In recent times, industrial nations have felt threatened by the industrial developments in poorer countries. As protectionist thinking surfaces, as Burke-Hartke solutions that would "bottle-up" technology to preserve United States employment get hearings, the multinational corporation is not passive. MNCs are opposed to Burke-Hartke-type legislation. Their opposition results in a joined interest with LDCs. Like it or not, the less developed countries have a good ally in the multinational corporations.

NOTES

(1) Michel A. Amsalem, "Technology Choice in Developing Countries: The Impact of Differences in Factor Costs," unpublished D.B.A. dissertation, Graduate School of Business Administration, Harvard University, 1978. Amsalem makes it very clear how many different parts of a production process can have alternative technologies.

(2) For a start, see Yair Abaroni, The Foreign Investment Decision Process (Boston: Graduate School of Business Administration, Harvard University, 1966); Ray Vernon, Sovereignty at Bay (New York: Basic Books, 1971); Mira Wilkins, The Emergence of Multinational Enterprise (Cambridge, Mass.: Harvard University Press, 1970); Mira Wilkins, The Maturing of Multinational Enterprise (Cambridge, Mass.: Harvard University Press, 1974).

(3) See for example Jack Baranson, "Technology Transfer Through the International Firm," American Economic Review, 60 (May 1970): 435-40.

(4) The quotation is from Economic Development Laboratory, Employment Generation Through Stimulation of Small Scale Industry, Atlanta, 1976, p. 1. The context was "selected aspects of intermediate technology." We have taken the quote out of context.

(5) W.J. Baumol, Business Behavior: Values and Growth (New York: Basic Books, 1959) (sales); Robin Marris, The Economic Theory of Managerial Capitalism (New York: Basic Books, 1964) (growth); R.M. Cyert and J.C. March, A Behavioral Theory of the Firm (Englewood Cliffs: Prentice-Hall, 1963) (satisficing). For various other "theories" of multinational corporations, see Charles P. Kindleberger, American

Business Abroad (New Haven: Yale University Press, 1969); Charles P. Kindleberger, The International Corporation (Cambridge, Mass.: M.I.T. Press, 1970); Louis T. Wells, The Product of Life Cycle and International Trade (Boston: Graduate School of Business Administration, Harvard University, 1972).

(6) We have discussed this at length in Mira Wilkins, "The Role of Private Business in the International Diffusion of Technology," Journal of Economic History, 34 (March 1974): 168-172.

(7) See Vernon, Sovereignty at Bay.

(8) For example, Alexander Gerschenkron, Economic Backwardness in Historical Perspective (New York: Praeger, 1965), pp. 9, 26.

(9) After writing most of this, I read Professor Miguel Wionczek's useful paper, "Less Developed Countries and Multinational Corporations: The Conflict About Technology Transfer and the Major Negotiable Issues," U.N. Center for Transnational Corporations, New York, December 1977. This view corresponds in some ways with what he calls the "orthodox" view, although my group of "yes" responders – unlike his "orthodox" group – does not go into the question of internal transfers within the multinational corporation or the problems of unbundling – or "unpacking" as Wionczek calls it. It assumes that the multinational corporation can transfer its whole package or a part that is enriched by association with the whole. I added the last sentence after reading Dr. Wionczek's paper.

(10) The quoted words are from Dr. Wionczek's paper (see note 9). Here again, with the exception of the last sentence, I wrote this before reading his paper.

(11) This point is made in Louis T. Wells, Jr., "Economic Man and Engineering Man: Choice of Technology in a Low Wage Country," Public Policy, 21 (Summer 1973).

(12) Amsalem, op. cit., in his dissertation deals only with existing, available technologies. He shows clearly that the number of alternatives varies by stage of the process.

(13) Wionczek, op. cit., pp. 10, 17.

(14) Dimitri Germidi, ed., Transfer of Technology by Multinational Corporations, 2 volumes (Paris: OECD, 1977), I, p. 26, is so fearful of multinational corporations that it warns developing countries to "be very careful when calling for the decentralization of R&D by the MNC. If the product of research does not directly and amply benefit the local, national, or regional economy, the small scientific potential of the host country is going to contribute to the scientific and technologi-

cal development of the source country, to the detriment of its own advantage."

(15) Ibid., I, p. 28, concedes that local purchases by multinational enterprises are a "possible way of stepping up the local technical and industrial potential" – yet, despite substantial evidence of this in the case studies that follow – the report is so hostile to multinational enterprises that it goes overboard with reservations. (That multinational corporations require technical standards is, for example, seen not as a desirable upgrading of national technology, but as an exercise in unwarranted control!)

(16) This type of aid to suppliers and dealers occurs in both developed and less developed countries. See Wilkins, "The Role of Private Business," p. 178, especially note 40.

(17) Wionczek, op. cit., p. 28.

(18) D. Babatunde Thomas, "An Economic Theory of International Technology Flows and Productivity Differences." Paper presented at the Fifth World Congress of Economists, Tokyo, Japan, September 1977.

(19) Amsalem, op. cit., p. 174, found that in his limited sample of government-owned firms in less developed countries that they were more capital-intensive than the private firms studied. "Cost minimization . . . played a very limited role, if any, in their choice of technology, while criteria of engineering excellence and risk minimization seems to have the largest weight in their decisions." The firms Amsalem studied were managed by engineers "with limited business experience" and, as far as he could determine, "the choice of technology was made on technical rather than economic grounds."

(20) This point is made by Amsalem, op. cit.

(21) Wionczek, op. cit.

(22) P.S. Lokanathan, Industrial Organization in India (London, 1970), pp. 15-16; Vera Anstey, The Economic Development of India, 4th ed. (London, 1952), pp. 113-15,591-95.

(23) The notion of the importance of both imitation and innovation comes from Stuart Bruchey, The Roots of American Economic Growth (New York: Harper and Row, 1965).

6 Do Multinational Corporations Really Transfer Technology?

Trevor M.A. Farrell

The assertion that multinational corporations (MNCs) transfer technology to LDCs is widespread in the literature on technology, foreign capital, and economic development – both academic and nonacademic. One of the most frequent arguments offered in defense of the MNCs is that they are the major, most effective, and even the only, channel through which LDCs can expect to obtain the technology they need.

Widespread as this claim is, it is almost never examined, and one searches hard, long, and usually fruitlessly, to find the claim backed-up by hard evidence on the part of those who make it. It is treated as obvious, even axiomatic by some writers and spokesmen. It is treated as an article of faith by others. It is not treated as a scientific proposition to be precisely defined, examined, and investigated.(1)

The fact is, however, that the situation is not so straightforward at all. Foreign metropolitan capital has been operating in the LDCs generally for at least a century. For some of these countries, the Caribbean for example, foreign capital has operated for two centuries or more. Even in its specific, modernday manifestation as the MNC, LDCs have been hosting this form of foreign capital for several decades. The oil industry is the most notable example. In Trinindad-Tobago for example, MNCs have been operating and controlling the local oil industry for over sixty years. In the Middle East, the MNCs' involvement in oil goes back anywhere from forty to seventy years. Yet, in all these cases, the requisite indigenous technology for operating the industry effectively and totally still is held not to exist.

One can think of several examples where, after the operation of MNCs for several decades, a country nationalizes an industry and then discovers it cannot run it effectively. But, presumably, technology was transferred. Or was it? How can we reconcile the confident certainty, so widely expressed, that MNCs transfer technology, with the objective fact that LDCs remain technologically dependent, underdeveloped, and incapable of operating their own industries, even after decades of MNC

69

involvement? It is high time for a serious examination of the supposed transfer of technology. Is technology, in fact, transferred? What technology is transferred? To whom? What is meant by 'technology'? What is meant by 'transfer'?

TECHNOLOGY AND ITS TRANSFER: AN OVERVIEW

This chapter seeks, first of all, to examine in detail what is meant by technology and its transfer. Then, it attempts to pinpoint some of the reasons why the popular, and casually advanced, proposition that MNCs transfer technology needs to be carefully qualified, if not rejected.

First of all, what do we mean by the term 'technology'? Technology may be defined as the methods, processes, and techniques used in the production of goods and services. This might be termed production technology. Consumption technology, on the other hand, refers to the methods, processes, and techniques by which a particular need or demand may be satisfied. For example, the need for internal transport may be satisfied through the use of horses and buggies, automobiles, subway systems, trains, buses, or bicycles. The word 'technology' in popular usage generally refers to production technology and that is what we are primarily concerned with here, and what we intend the unprefixed work 'technology' to mean.

Now, technology refers not only to mechanical or technical aspects of production, but also accounting and inventory systems, personnel management techniques, organizational systems, and so forth. Therefore, the term technology or production technology has to be interpreted broadly. Further, the essence of technology is knowledge. In the modern world, technology really means the systematic application of knowledge, derived from the scientific study of the physical world and its properties on the one hand, and human experience on the other, to the solution of problems.

Technology is knowledge – knowledge repositing in the human being and combined with understanding. Technology, therefore, is not machinery and equipment. The highly visible, physical artifacts of technology (jet aircraft, computers, semi-submersibles, radar), which we so often identify as being synonymous with technology, are, in fact, nothing more than the embodiment of technology.

Immediately, we can identify one source of the automatic presumption that MNCs transfer technology. The necessity to locate often highly elaborate and very visible artifacts of production in LDCs (e.g. oil refineries, earth-moving equipment, etc.) will lead to the belief that technology has been transferred, as long as we mistakenly identify technology with physical artifacts. The invention, design, and construction of any piece of equipment, or any production process is utterly dependent on human knowledge and human skill. Furthermore, without people, any physical artifact, immediately or ultimately, will mean little or nothing. The question of technology, then, is really the issue of human resources – human beings and their knowledge, skills, and capabilities.

Detailed study of the actual operations of several industries suggests that, in considering the technology needed to effect an operation, it is necessary to analyze the concept of production technology in greater detail. Two points that emerge are pertinent to our discussion here. The first is that in almost every industry, successful operation requires not one technology, but the conjoint application of several. For example, study of a multisectoral oil industry in Trinidad-Tobago – one involving exploration, production, and refining – showed that sixteen different basic functions could be identified. These, when further sub-divided, ran into well over fifty different areas of knowledge or skill.(2)

To run an industry or operation over time will require the carrying out of many different functions, requiring many different technologies. To run it even for a short period requires the conduct of a certain minimum subset of these functions. The functions in this subset can be defined as the critical skills. To talk about the transfer of technology, then, and the developed ability of nationals to carry out an operation, we often have to focus attention not on one technology but on a complex of interrelated technologies.

The second point that is important for our purposes here is that there are different 'levels' of knowledge relevant to a production process, and these different levels have different levels of significance for the successful carrying out of operations. We can make another basic distinction between two kinds of technology – static and dynamic.

Static technology refers to those kinds of knowledge which, once possessed, permit the successful carrying out of certain routine operations, in a more or less fixed fashion, and with more or less given equipment. The skills of the typist, the bookkeeper, the refinery operator, the welder, and the PBX operator are cases in point.

By contrast, the possessor of dynamic technology usually comprehends the scientific principles and other considerations undergirding his work.   As such, he tends to have the potential for invention and innovation. He possesses an overview of his function and understanding of the fundamental bases on which it rests. For example, the bookkeeper ordinarily cannot compete with the accountant in designing a new, complex accounting system. The nurse cannot usually supplant the research doctor in devising and testing new techniques of treatment for some disease.

Naturally, the distinction between static and dynamic technology, while real enough, is not hard and fast. The one shades into the other. Also, the possessor of static technology is not absolutely precluded from innovation or, more precisely, from innovative ideas. In industry, many new ideas come first from workers on the shop floor. Usually, though, their translation into reality depends upon the assistance of the scientists and engineers. Similarly, not every possessor of the kind of training usually associated with dynamic technology will, in fact, turn out to be an innovator of any worth or consequence.

The significance of the distinction between static and dynamic technology lies in its implication for the continuity of an operation or industry over time. Successful and continuous functioning requires that an organization prove able to identify and solve the new problems which beset it in a changing environment. For example, an oil company which was engaged in land production may find that new reserves lie offshore, perhaps in difficult and even hostile environments (say, the polar caps). It needs to find the techniques and processes which will enable it to solve an altered problem set in order to stay in existence. This is where dynamic technology comes in.

There have been several experiences where ordinary workers have run plants for a while after the managerial and technical staff have left. Such enterprises, manned by staff who have been well trained to perform certain limited, routine functions may continue to operate for some time on a limited basis, like a car coasting downhill with the engine off. But eventually, as new and/or major problems arise, which require a conceptualization of certain fundamental scientific principles, such enterprises tend to face increasing difficulty and even disintegration, unless the necessary dynamic technology is acquired.

THE MEANING OF TECHNOLOGY TRANSFER

The notion of 'transfer of technology' has to be understood in terms of these several complexities. It is only meaningful to talk about the transfer of technology to poor countries in terms of some specific operation of industry or economic activity. From a developmental perspective, the value of technology is primarily the contribution it makes to production and the alleviation of poverty. For a technology transfer to be translated into production, it must, at some stage, directly or indirectly, be incorporated into some actual activity.

Since, as pointed out above, virtually any modern economic activity requires a range of technologies for its operation, the contribution of a transfer has to be seen in relation to the range, or set, of technologies required for the operation.

Carrying out an activity, or operating an industry, has to be discussed in some timeframe. The true conferral of the ability to run some industry implies the ability to run it over time. This means acquiring the capability to identify and solve the changing problems which arise. This, in turn, means the acquisition of the relevant dynamic technologies.

The international transfer of technology implies the transfer of certain capabilities from one country to another – and, specifically, to the citizens of the LDC and to its 'true' residents. The term 'true residents' is used here as opposed to, say, expatriates who may, though essentially temporary, have acquired legal resident status.

Finally, the transfer of technology across national frontiers by MNCs implies that 'attribution' should be possible. That is, it should be possible to link the genesis or cultivation of some skill(s) directly to the activities of the MNCs. Direct linkage means that the imparting of the particular technology should be through the training or other tutelary activities of the company.

It is now possible to essay a more precise definition of what the term 'transfer of technology' should be taken to mean. The international transfer of technology by MNCs is really the development of ability on the part of nationals of the receiving country, as a direct result of the corporation's activities, to operate a given industry or activity successfully over time.   Successful operation over time implies the development of a problem solving capability.

THE MNC AND THE TRANSFER OF TECHNOLOGY TO LDCs: AN ASSESSMENT

The preceding discussion now makes it possible to identify and illuminate several issues which qualify, and in some cases even negate, the idea that an automatic, beneficial, and productive transfer of technology arises out of the relationship between MNCs and LDCs.

The relationship between MNCs and LDCs today takes two major forms.   On the one hand, there is the traditional form of direct investment, with the operation of a wholly or partly-owned subsidiary in the host country.   On the other hand, the relationship may take the form of some sort of technology arrangement (e.g. licensing, marketing, or management agreements). In analyzing the transfer issue, it is necessary to deal with both types of relationships.   There are eight points that emerge from this analysis and from empirical research work, which qualify, or negate, the customary assumptions about MNC technology transfer and its effects.

One may find the phenomenon of a MNC operating a wholly-owned subsidiary in an LDC for several decades and, despite the fact that locals occupy 90 or 95 or even 99 percent of the positions in the enterprise, they prove unable to operate it when the MNC pulls out, or is forced out. Why? One simple reason is that the complex of functions necessary to successfully operate the enterprise or industry over time was never carried out entirely in the host country.

For example, detailed study of the functions carried out in the Trinidad-Tobago petroleum industry (which is, and always has been, dominated by foreign MNCs), and a comparison of these with the entire complex of functions needed to successfully operate an oil industry, revealed that several functions were never carried out locally at all. For example, 1) 'Major Projects – their Planning, Design and Construction', 2) R&D, 3) economic analysis and planning, 4) international marketing, and 5) marine tanker transportation were almost totally absent.(3) Other functions, though nominally carried out, existed at a low level.

Therefore, it is clear in these cases that where no experience with certain functions exists, and no technologies have been developed in these areas by nationals, the LDC will be unable to operate the enterprise or industry on its own. If the missing functions include critical skills, then it may not even be able to operate the enterprise in the short run without outside help. In several cases, what seems to happen is that the kinds of functions which are not performed locally, and where there is not even the semblance of transfer, are what may be labelled the 'cephalous' or 'brain' functions. Examples include planning and R&D.

Another reason why transfer, as defined above, may not take place lies in the selectivity with respect to the level at which technologies are, in fact, imparted through MNC operation. The successful conduct of many functions requires a coordination of static and dynamic technology. For example, the conduct of the accounting function in an enterprise may require both accountants and accounting assistants, clerks or bookkeepers. The conduct of an engineering function may require the coordination of design engineers and draftsmen, or operations engineers and technicians.

The multinational corporation, by reason of its geographical spread across national frontiers plus modern communications, is able to split functions internationally. It may locate the areas that are the province of static technology in the host underdeveloped country, while those that are the province of dynamic technology may be retained abroad. Modern communication permits the successful coordination of the particular function, which requires that both levels of technology complement one another across wide geographical space.

Thus, companies may train accounting assistants, bookkeepers, computer programers, welders, machine shop operators, and others locally. But the accounting, computer science, design engineering which are relevant to the local operations – i.e. the relevant dynamic technology – may take place abroad. Often, then, when observers point to MNCs transferring technology, all they are really talking about is static technology.

It is important, too, to point out that the MNC may even relegate professionals to the performance of the routine functions associated with static technology. Thus, professional accountants or engineers, despite their certification, their salaries, their scientific training, and the glorified titles on their office doors, may be relegated to functioning as little more than bookkeepers or technicians.

The LDC accountants may never be called upon to participate in the design of the accounting system. Their job, instead, is simply to apply the procedure set out in multivolume, company accounting manuals. The LDC engineers may never participate in the selection of production processes, or in the design or choice of equipment. Instead, their job may be simply to run given processes and equipment according to prescribed guidelines.

The host country may then discover, on nationalization day, that its nationals – despite their titles and decades of 'experience' with the company – are fundamentally incapable of successfully running the operation, identifying and solving its problems. They may have never been really exposed to the areas of dynamic technology, their professional skills and scientific ability acquired through formal education may have atrophied from disuse, and, most important of all, their values and attitudes formed through years of dependence may render them incompetent and unable as a bird who, though possessed of wings, has been caged all its life.

Another frequent error is the identification of the location of equipment and sophisticated processes in LDCs with the transfer of technology. This is compounded when the static technology required to operate that precise set of equipment under highly specific conditions can be seen to have been developed. This is the deceptive allure of the turnkey plant, for instance.

It is time to consider, now, more precisely the transfer issue as it arises through technological arrangements concluded between LDC enterprises or their governments on the one hand, and foreign MNCs on the other. This is a significant and rapidly growing arena for the supposed transfer of technology. These arrangements are usually effected through the medium of licensing agreements, management contracts, service contracts, or sales agreements.

It has been pointed out by Vaitsos that the term 'transfer' is clearly inappropriate.(4) What is involved is a market transaction in which a price is paid and, presumably, a sale effected. It is necessary to note, however, that this is a curious kind of sale for several reasons. Unlike the usual commodity, purchase of a technology does not mean exclusive appropriation by the buyer with concomitant rights to exclusivity in possession, and with the power of subsequent disposal at his discretion.

Knowledge sold remains fully in the possession of the seller. Licensing agreements also usually carry a nonexclusivity clause, so that the seller may grant multiple licenses of the same knowledge. Also, the use of the knowledge by the licensee is often restricted and, through clauses forbidding sub-licensing or demanding secrecy, he cannot diffuse it to others.

What we have, then, is not the transfer of technology, but its commercialization. Furthermore, it is often inappropriate to think of commercialization as meaning sale. What we have is more nearly the renting or leasing of technology.

Research on the vehicle assembly and textile industries of Trinidad-Tobago again demonstrates the point.(5) The licensing agreements in both these industries prevent diffusion by the buyers, do not confer exclusive rights, and do not permit further use of the knowledge involved after the expiration of the agreement. The term 'transfer' is clearly a misnomer.

The study of actual technological agreements between MNCs and LDCs brings up another very important question with respect to technology transfer. Exactly what technologies are the subject of these agreements? Research into the operation of several industries or other economic activities demonstrates that the successful conduct of an activity is generally dependent upon certain key technologies. The ability to carry out the particular activity is ultimately dependent upon a transferee's acquisition of these key technologies. But these key technologies are, in several cases, not the subject of technological agreements at all. An LDC operating from a position of ignorance, and concluding or sanctioning an agreement it believes will ultimately lead to its development of a technological capability in some industry or activity, is often deluding itself. What it may, in fact, be acquiring are peripheral technologies, the possession of which, without the key technologies, will never permit it to carry out the particular operation on its own. An example may be taken from the automobile industry.

Modern automibile manufacturers have three key technologies which underlie their success. These are:

1. A capacity for design and product innovation;

2. The expertise for the efficient search and selection of parts, materials, and components;

3. The expertise for efficient general management, particularly in such areas as production scheduling, plant layout, and inventory control.

LDCs which enter licensing agreements with foreign manufacturers to assemble vehicles locally are often not obtaining the technologies which would permit them one day to shake off their dependence. What they may end up doing is leasing the technology for assembling vehicles. This technology is essentially secondary, and in any event not mysterious. And even here, the skills provided may be simply static technology. The Trinidad-Tobago vehicle assembly industry is, again, a clear example of this type of situation.(6)

Undoubtedly, many technological arrangements between MNCs and LDCs do not provide the recipient country with the key dynamic technologies necessary for a true transfer as defined above.

A related phenomenon is where MNCs provide old, outdated technologies, either just about to be superseded, or already superseded by new, more advanced technologies. LDCs then set up industries which would never be capable of competing effectively on international markets against their erstwhile mentors. All this is perfectly sensible from the point of view of the MNCs. At the same time, it is certainly not in the interests of the poor countries involved. It is, in fact, a simple continuation of the old trick of the white man: selling technologically outdated muskets to the black man or the red man at exorbitant prices, while making sure to retain the advanced firearm to outgun him when the time comes.

The issue of what is transferred or commercialized also arises in another context — the relevance of the particular technology to the LDC in question and its functionality or dysfunctionality. The issue of technological functionality/dysfunctionality arises with respect to the appropriateness of a particular technology. The technology for producing supersonic aircraft, or for snow clearance, refining residual fuel oil, or producing polyester doubleknits may all be irrelevant to the real needs of a tropical country in the Caribbean or Latin America. Also, highly capital-intensive technologies may be singularly inappropriate to a labor-surplus country experiencing unemployment, when cheaper, more labor-intensive techniques are available or can be developed.

Therefore, in analyzing the technological relationships between LDCs and the MNCs, it is important to consider the relevance and functionality of whatever technology is at issue. Even if a genuine transfer of technology were taking place in a given case, it is necessary to go further and examine exactly what is being transferred.

Another obvious, but generally ignored, problem that arises in discussing the impact of technological arrangements is the question of attribution. MNC subsidiaries operating in underdeveloped host countries may do quite a lot of training. Licensing agreements may prescribe training for locals in licensor plants or provide for training visits by licensor technicians to the licensee's operations. But it is not the case that all the skills, static and dynamic, possessed by nationals operating in the particular concern are attributable to this relationship.

In many cases, nationals in a country will proceed on their own initiative to acquire skills relevant to a particular operation. The existence of the industry or activity acts as a catalyst for them to do so, since they may hope to later gain employment. One can hardly attribute the acquisition of such education and skills, undertaken as an investment in future employment, to be the direct responsibility of foreign concerns.

Finally, one of the clearest reasons why no transfer of technology may take place, whether through direct investment or through arms-length technology agreements, is that the necessary preconditions for the successful reception and absorption of technology in the host country may be absent. While it is possible to identify these conditions, this intriguing issue cannot be pursued within the confines of this chapter.

CONCLUSION

For all these reasons, then, it is quite clear that the automatic presumption that MNCs, in fact, serve as a channel for the transfer of technology to LDCs is unjustified. The term transfer is itself misleading and inappropriate. Multinational corporations may provide some, but not all, of the requisite technology to operate an activity. They may fail to provide the relevant dynamic technology, or any of the key technologies. What is, in fact, transferred may be fundamentally irrelevant or inappropriate. And it may be a mistake automatically to attribute to these companies all of the technologies possessed by locals operating in a given activity. The much vaunted transfer of technology by MNCs turns out, on examination, to encompass considerable illusion – at least on the 'South' side of the North-South divide.

NOTES

(1) Examples chosen at random include U.N. (United Nations), Multinational Corporations in World Development (New York: U.N., 1973), U.N. (United Nations), The Acquisition of Technology from Multinational Corporations by Developing Countries (New York: U.N. 1974), A.K. Cairncross, "The International Transfer of Technology," in James Theberge (ed.), Economics of Trade and Development (New York: John Wiley & Sons, 1968). But literally dozens of examples can be found.

(2) Trevor M.A. Farrell, "The Multinational Corporations, the Transfer

of Technology and the Human Resource Problem in the Trinidad-Tobago Petroleum Industry," (mimeo) Caribbean Technology Policy Studies – (CTPS) 1977.

(3) Ibid., p. 48.

(4) Constantine Vaitsos, "The Process of Commercialization of Technology in the Andean Pact," in Hugo Radice, ed., International Firms and Modern Imperialism (Baltimore: Penguin, 1975).

(5) Trevor M.A. Farrell and A.M. Gajraj, Technology and the Manufacturing Sector in Trinidad-Tobago: The Textile/Garment Industry, (mimeo) C.T.P.S. 1977. Trevor M.A. Farrell and A.M. Gajraj, Technology and the Manufacturing Sector in Trinidad-Tobago: The Vehicle Assembly Industry, (mimeo) Caribbean Technology Policy Studies 1977.

(6) Ibid.

III
Technological Problems in the Caribbean

7 Intermediate Technology and the Modernization of Anglophone Caribbean States

J. Dellimore

About a decade and a half ago, the decolonization of the Anglophone Caribbean began. In the post-independence period, government intervention has come to be seen as the key instrument for accelerating national development. Informed by theories based on a top-down development process, governments of the newly emergent nations have attempted to launch their societies on a path of accelerated social and economic advancement by adopting strategies based on: a mixed economy with centralized planning and decision making; rapid modernization, through expansion of the existing modern sector, industrialization and a massive inflow of foreign capital and advanced technology; and the fullest participation of the national elite in the expansion and diversification of economic activity. Preambles to national development plans provide ample illustrations of this development philosophy. Within this framework the starting point of government strategies has been the demands and interests of foreign investors and the national bourgeoisie. To encourage their participation in national economic activity, there has been a proliferation of inducements to such groups in the form of tax concessions and subsidies (mainly in the provision of overheads such as education, roads, public utilities, port facilities, factory space at pepper-corn rents, etc.).

There is little doubt that government intervention has, so far, failed to bring about the social and economic advancement anticipated fifteen years ago. While there has been some economic growth, significant improvements in public health standards, and a substantial expansion in their education systems, a significant restructuring of the economic and social systems of these states has not occurred. In fact, in many crucial aspects of development, the problems have been worsening, and there have been growing contradictions between the avowed purpose of government intervention and the consequences of such intervention. Thus, unemployment levels are rising and now exceed 25 percent in many of these states and social divisions and rivalries have been intensifying. An increasing proportion of strategic basic needs (particularly food) have to be met by imports, there is a growing incidence of malnutrition and undernutrition (after an initial decline), and there is increasing subordination of national interests to those of metropolitan powers in the way the economies operate.

Nowhere are the failures of existing policies more apparent than in the use of technology. As a creator of dependence and social inequalities, technology has emerged as being equally or more important than capital. In the absence of an effective indigenous, scientific and technological capability, foreign technology is being integrated into national economies largely on its own terms, so that the unfavorable international distribution of the benefits from, and control of, national production established in colonial times has persisted in independence. Consumption habits are being promoted, and development is being encouraged and organized along lines which lead to demands for goods and services requiring technologies which are accessible only to large investors, thus denying the lowest socioeconomic groups opportunities to exercise any significant economic initiative.

As efforts to accelerate industrialization increase, contradictions in technology policy have been intensifying. Thus, in these supposedly capital-scarce economies, there is massive underutilization of plant capacities and an ever-increasing outflow of capital as payments for licenses and other technology rentals used for purposes of dubious value to our societies. Production relations embodied by imported technologies, or encouraged under the system of incentives, have the effect of restricting rather than releasing the energy and enterprise of low income groups who form the bulk of the population. Today the mass of people are increasingly being seen as a burden or constraint on development, rather than as a valuable asset. Perhaps the greatest contradiction of all is the rising demand for the 'transfer' of advanced foreign technologies, despite the general experience that such transfers require that governments make more intimate, and continue to consummate, relationships with the same foreign industrial powers to whom most of the forces maintaining the underdevelopment of these states are attributed.

The challenge to science and technology policy for the Caribbean is the modernization of the region (that is, the creation of a Caribbean Community which is completely adjusted to its time and capable of a total response to internal and external changes under indigenous direction) by appropriate and effective combinations of available scientific, technological, material, and human resources. Among the major obstacles to the development of such a situation is the fact that, within the existing dependence framework, virtually all formal instruments for implementing policy are ineffectual. One Caribbean economist has noted, for example, that if the various policy implementing and regulating institutions were closed down, it would have little or no real effect on what actually takes place.(1) Existing regulatory institutions are mere symbols of independence. Evidently, in any alternative policy, greater emphasis must be placed on the operation of informal instruments of policy and on the creation of the preconditions for formal policy instruments to become effective.

The process of modernization will, to a large extent, involve breaking away from old habits and practices and adopting new ones, so as to optimize our benefits from a changing world. The ultimate purpose of attempts to bring about such progressive social, economic, and political changes must be the development of Caribbean people so that both collectively and individually we may realize our full potential. Collective development and individual development could be regarded as the two lines along which Caribbean societies should seek to advance through modernization.

Our historical experience shows that the basic minimum of achievements which could launch individuals on a path leading to the unfolding of their potentialities are:

1. Satisfaction of basic needs for food, shelter, clothing, health, education, and community.

2. Access to opportunities for self-development, particularly through gainful employment above subsistence level and through participation in decisions affecting their welfare.

3. A feeling of a distinct identity, self-confidence, and a creative potential.

4. An ability to face the world with purpose and pride.

Corresponding minimum achievements, at the community level, if our societies are to be launched on a path of collective development, are:

1. Development of a productive system capable of meeting the basic needs and reasonable aspirations and demands of all individuals within the collective.

2. Self-reliance based on faith in the collective potential and principles of self-determination.

3. A vision of a collective destiny shared by the vast majority of individuals in the society.

4. Maximization of collective welfare through optimal use of available resources, of which labor is the most important part.

The above should be regarded as defining a framework of objectives and informal policy instruments within which, and by means of which, planners should seek to reformulate science and technology policy for the Anglophone Caribbean, if they wish to bring about the fullest absorption of all social groups into modern sectors of our economies.

A crucial aspect of any effort by governments to accelerate and coordinate advancement along these two lines is the way science and technology are used in promoting these objectives. In particular, the behavior of technology should be vigorously controlled insofar as it affects relations between foreign powers, the state, and the different competing social groups within the state. Thus, it will be imperative that the modernization of our economies be primarily internally driven – not in the sense of starting from scratch to reinvent or reproduce what has been done elsewhere, but in the sense of understanding, mastering, and directing the course of changes free from foreign domination. It is equally imperative that there be massive utilization of what is now regarded as surplus labor, under production relations which allow every citizen the opportunity to feel they are sharing in, and contributing to, the advancement of their society. Another important requisite is the creation of conditions in which the lowest socioeconomic groups can take some initiative in efforts to improve their lot, thus giving them the power to protect their interests against predatory social groups from within or without their society.

INTERMEDIATE TECHNOLOGY: ITS SIGNIFICANCE FOR SMALL, UNDERDEVELOPED DEPENDENT SOCIETIES

The concept of intermediate technology and its implications for development are best understood in terms of Schumacher's 'Law of the disappearing middle.'(2) According to this law, if one views technological development as occurring in stages so that the most primitive type represents level I technology, then as changes come about which take basic technology through successive development levels, as each new level is realized, technologies corresponding to all levels between the first and the most advanced tend to disappear. Thus, at any stage in the evolution of technology, only the most primitive and the most advanced technologies tend to coexist in a society. This would seem to be particularly true of underdeveloped, open economies such as are found in the Caribbean. In practice, though not always, one finds that levels of technology as defined by the 'Law of the disappearing middle' correspond roughly to different levels of equipment cost, centralization, local raw material content, production scale, labor intensiveness, and complexity. On this basis, the term intermediate technology has come to be applied to technologies which are:

1. Low-cost (especially in relation to the average yearly earnings of workers and peasants).

2. Users of materials (and skills) which are predominantly local in origin;

3. Small-scale;

4. Users of decentralized (and, ideally, renewable) energy resources;

5. Labor-demanding and skill-developing;

6. Low-level in complexity; and

7. Local community oriented.

The 'Law of the disappearing middle' does more than explain the origin of the term intermediate technology; it is, in fact, a statement of a principle of dualism. The classic example of dualism in the Anglophone Caribbean is said to be the agricultural sector. In this sector small peasant farmers engage in mixed subsistence and commercial farming on small plots of land, using primitive implements such as the hoe, fork, and cutlass for tillage and reaping, primitive methods of soil management based on pen-manure, and slash and burn methods for clearing land, – all of which are centuries-old techniques dating back to the time of settlement. Within the same economy and, often within the same district, one can find farms and plantations using advanced techniques such as mechanized tillage and harvesting; extensive application of fertilizers, herbicides and pesticides; support by ongoing R&D; commerical credit; etc. This coexistence of primitive and sophisticated production techniques and forms of economic organization is now a dominant characteristic of the agricultural sector.

But dualism is not restricted to agriculture, it exists in banking and finance, in fisheries, in the furniture industry, in the garment industry, and in the food processing industry. Indeed, almost every sector of Anglophone Caribbean economies (with the possible exception of mineral extractive industries, and certain public utilities such as electricity and telephone services) has a technically sophisticated, high productivity, typically outward-looking branch, and a primitive, low productivity, typically community-oriented branch. This dualism and its attendant consequences (particularly the marginalization of large sections of the small societies) is one of the important, if not the most important, immediate causes of the economic and social problems which present obstacles to the rapid development of these states. Modernization has failed to develop throughout their economic systems because major sections are effectively insulated from modern techniques by the mechanisms maintaining, or associated with, dualism. The 'Law of the disappearing middle' posits the technological dimension of the dual economy.

Somewhat paradoxically, there are good reasons for believing that the processes by which intermediate technologies are made to disappear in these underdeveloped economies form a major part of the indigenously-generated component of the barrier to modernization and technical advancement. There is evidence that, under existing policies, the disappearance of intermediate technology from some manufacturing activities is very closely linked with efforts by the larger private investors to establish oligopolistic market structures (in which they can operate in collusion or as monopolists), rather than with desirable and appropriate improvements in production or consumption efficiency. These larger investors draw heavily on outside technical support to win and maintain considerable advantages over intermediate operators by successfully promoting the adoption of sophisticated foreign consumption habits which require production technology out of reach of, and grossly inefficient for, intermediate operators. Both the demand for technology and, consequently, the demand for raw materials come to be increasingly directed to external sources as development proceeds along these lines; and direct linkages between the local resource base and local consumption aspirations diminish with the unregulated elimination of intermediate operators who are more likely to make intensive use of indigenous resources and cater to traditional tastes.

The main weakness of intermediate operators is their lack of dynamism; their main strength is a favorable resource and consumption orientation. Users of advanced technology are no more dynamic in relaton to the stimulation of the local economy, though more open to change, and they are much worse in their resource and consumption orientation. One needs to find ways to advance technologically while preserving the most favorable attributes of intermediate operators and imparting dynamism to one's economy.

The queston arises: can support for existing intermediate operators, or a selective introduction of intermediate technologies, in Caribbean mini-states be used to accelerate the pace of modernization by reversing the processes by which large groups of people are entrapped in backward branches of productive sectors under conditions in which there is a steady worsening of their economic and social positions relative to other groups in their society? It is worth noting that the conditions of employment of wage labor associated with 'intermediate level' operators are generally much worse than for wage labor associated with operators of advanced technology within the prevailing framework of exploitation of the working class. Antecedent ideological changes will be crucial to the success of any strategy to accelerate development using intermediate technology; if such changes do not take place, this strategy could well worsen the position of low income groups and create a strong petit-bourgeoisie. The basic proposition of this paper is that if acquisitions of advanced technology are to be used to stimulate the spread of modernization throughout our economies and are not to reinforce dualism, measures (or steps) to acquire advanced technology should, as far as possible, be linked with measures (or steps) to acquire, disseminate, or support complementary intermediate technologies which can have the effect of drawing backward and modern operators closer together – in terms of their interests, their ability to participate in the economy, and the share of benefits they derive from the economy.

Three strategies can be adopted in the deployment and support of intermediate technology in manufacturing industries.

1. The pilot plant strategy. The Peoples' Republic of China is said to have used this approach to deal with a scarcity of trained manpower and suitable raw materials in the establishment of 'local' industries.

2. The subcontracting strategy believed to have played a significant part in the modernization of Japan and the Peoples' Republic of China.

3. The optimal-spread strategy.

These strategies should not be seen as mutually exclusive, but complementary, so that some projects may combine all three.

The Pilot Plant Strategy

The objective of the pilot plant strategy should be to facilitate diversification of our economies and greatly increase mass participation in finding ways to develop our resources and in solving problems of underdevelopment. Within this framework, intermediate technologies can be used in the search for:

— alternative uses for by-products, surpluses, and abundant resources; and for

— tapping small concentrations of energy, raw materials, or small market opportunities, as a prelude to more massive investment.

The main advantage of using intermediate technology for this purpose will be in allowing greater freedom for experimentation and exploration with minimum financial risks. The most vital aspect of intermediate technologies in such situations is their inherent flexibility and adaptability even in the hands of people with limited technical knowledge and skills. Thus, the rudimentary nature of most intermediate technologies and their lack of complexity can make them virtual pilot plants in the hands of our creative masses who invented the steel band and some of the most exotic rhythms in music to emerge this century, and are masters of one of the most creative action sports in the world.

Using intermediate technology, enterprising citizens with low cash incomes can respond to very small, or very uncertain, market or resource-processing opportunities with little fear of financial ruin because usually the same device can be adapted for use in a variety of activities. Some government support in the form of soft loans will be necessary, since the rate of capital formation by low income groups can be expected to be low if they are only able to draw on their own financial resources. The core components of an intermediate level sanding and scouring machine (used in shoe-manufacturing) can be used to drive a mill to shred coconut, or grind cassava, or macerate leaves, or to drive a lathe to make coconut jewelry. If any of the variety of applications take root, and markets and sources of raw material develop, the surplus earned and the experience gained in this process can be used to expand and extend production and modernize operations. In this way, the diversification and modernization of our economies can receive impetus from the efforts of enterprising groups drawn from the poor masses.

This pathway to modernization could also diminish existing foreign involvement in our economies. There are many reasons why such a reduction may be anticipated. First, sustained contact between the technology recipient and the original sources of intermediate technologies is usually not essential for the reproduction and assimilation of intermediate technology; the lower level of complexity of intermediate plant technologies means they are far easier to copy, adapt, and use than advanced technologies. Secondly, there should be a greater understanding and greater mastery of changes in the backward sector associated with the use of intermediate technology and, hence, greater scope for local decision making on the best use of the technology. Third, the low level of complexity and ease of assimilation of intermediate technology also means that it can be disseminated predominantly by our own nationals and be adequately supported by the weak, underdeveloped scientific and technological (S&T) systems of the region in ways which should encourage small industry and the S&T system to grow and develop together.

In support of the strategy outlined, it can be pointed out that there is no country in the region which has not had costly failures in using advanced technologies to pioneer new industries. Such failures have not only wasted scarce investment capital but have undermined national self-confidence and promoted even greater dependence on the judgment of foreign experts who are not without interests of their own. That foreign consultants indulge their own interests when proferring advice is evidenced, for example, by the fact that there is a strong tendency for consultants to recommend machinery from their country of origin.(3)

The Subcontracting Strategy

The objective of the subcontracting strategy should be to pull operators from the modern and backward branches of productive sectors together by encouraging operators with limited technical skills and financial resources to enter small linkage areas between established industries, creating a mode of organization of production in which operators drawn from the backward sector can be introduced to advanced production techniques under conditions which give them a greater sense of purpose and of their own creative and developmental potential.

Under this strategy modern enterprises would be encouraged to engage in cooperative production with small enterprises – for example, by sharing facilities for costly and scale-sensitive unit operations and processes, by the modern enterprise farming out to small enterprises operations which can be done competitively using intermediate technology (by persons who are not highly trained), and by governments regulating the sharing of markets by small and large operators. In situations such as these, by a gradual process of improving product standards, raising technical efficiency, increasing the small operators' share of responsibility for production and amalgamation of small enterprises, appropriate modern techniques could be introduced to backward operators.

The Optimal-Spread Strategy

The objective of the optimal-spread strategy should be to distribute the benefits of economic growth more equitably by opting for the development of goods and services at the level of technological sophistication which will enable more to be done with the little we have, and benefit the greatest possible proportion of the population. The much-debated question as to whether intermediate technology can be the rational economic choice of technique for these small island economies is the pivotal issue in the promotion and adoption of this strategy.

It is often said that intermediate technologies are less efficient than advanced technology in the use of capital; hence, we ought not to spread the limited, available investment capital over a large number of small operations which use intermediate technology; but should concentrate such capital on a much smaller number of more highly capitalized work places which use advanced technologies. A closely related belief is that, if industries were to be set up using intermediate technologies, we would merely be creating work for people, not creating viable business operations capable of surviving unaided in a competitive world.

There is no doubt that there is justification for such beliefs in some cases; no one could reasonably argue for the complete abandonment of advanced technology. The use of appropriate advanced technology is inevitable and quite desirable if the highest standard of living for the broad masses of people which our resources can support is to be achieved. However, the principle which underlies the optimal-spread strategy is the fact that in certain situations the use of intermediate technology can contribute more to the achievement of the preconditions for internally-driven, accelerated development, towards ultimate objectives, than advanced technology can.

Intermediate production technology can be quite competitive with advanced production technology if it is linked to intermediate consumption technology, and is operated in a setting in which there is only intermediate organizational technology. This is precisely the situation in backward branches of our productive sectors. Hence, in applying the optimal-spread strategy, it will be important that we do not try to produce, using intermediate technology, products whose specifications are exactly the same as those produced using advanced technology. It is equally important to this strategy that we do not restrict our vision of possible intermediate technologies to scaled-down versions of advanced technologies in which labor is substituted for machine operations. A search must be made for technologies which present distinctly different alternatives to advanced technology, that is, with fundamentally different resource and consumption orientations and different working principles.

If this can be done, it will be seen that, far from being generally uncompetitive and merely creating work for people, intermediate technologies can become overwhelmingly competitive in crucial development areas. For example, there is a growing acceptance that if basic health care is to be provided for every individual within the region then a paramedic will be needed to complement the work of highly trained doctors, who are expensive and hard to come by. A paramedic represents intermediate health care technology. Similarly, it should be possible for Caribbean people to become self-sufficient in protein, calories, and accessory food factors, without greatly increasing the land area under food crop cultivation, if Caribbean people can somehow be induced to adopt a less sophisticated consumption technology which is more efficient in the conversion of plant material into human food. For example, much protein is lost in converting plant material into human food by producing beef. Increased consumption of peas, beans, and other legumes could provide man's protein needs at a lower cost. Within a decade or two, it should be possible to provide adequate housing for every man, woman, and child if alternative intermediate-level building techniques and house designs, already partially developed, are adopted. For example, in Dominica, Richard Holloway has produced building blocks with adequate strength for house building, using as low as one-tenth the amount of cement as conventional blocks; his work has also shown that reinforcing steel and galvanized roofing could be replaced by much cheaper alternatives with some modifications in house building techniques.

A CASE STUDY – Problems and Prospects in the Development of the Food Processing Industry in the Eastern Caribbean Common Market (ECCM) States and Barbados(4)

It should be instructive to examine how the methodology and strategies for promoting the spread of modernization and industrial development outlined above can fit in with the situation on the ground in Anglophone mini-states of the Caribbean. For this purpose the problems and prospects in the development of a strategic subsector – the food processing industry – of the ECCM and Barbados are considered.

Nature and Characteristics of the Industry

The food processing industry is of greater importance to the economies of these states than any other manufacturing group. The industry has also shown a faster output growth since 1961 than the manufacturing sector as a whole. This expansion in output is largely due to the entry of new products into the industry. Thus, the output of established products has tended to remain constant over the years, with a few exceptions (notably animal feed production in Barbados). This stagnation in the output of established products, coupled with a substantial underutilization of plant capacity is largely responsible for the failure of governments to accelerate industrial growth through the stimulation of profits. Utilization of one-shift capacity is more often than not less than 50 percent and seldom exceeds 60-70 percent. Stagnation in the industry may, in turn, be linked to the fact that the industry is not needs-oriented, but is geared to producing items for which demand only grows slowly, as long as the basic needs of the bulk of the population are not met. Of the 77 products studied in the survey, 67.5 percent were luxury intensive, that is, products with a high degree of nonnutritional consumer satisfaction associated with their use and low nutrient cost ratios. Data for Barbados suggest that the industry has a more favorable ratio of employment to output than other manufacturing groups combined, and the average income of wage earners in the industry is well above the national average.

Dualism is not as marked in its extent in the food processing industry as it is in primary agriculture, and much of the dualism that exists takes on a peculiar feature. Thus, a high proportion of small food processors who use primitive techniques and who, typically, operate out of the home kitchen, are middle and low-middle income women. Many of these women have salaried jobs or husbands and family members with 'fair' incomes; their involvement in the industry is connected more with the survival of tastes for traditional products than with their own subsistence. Because of this, there is, typically, little interest in changing techniques or in expanding production in many 'backward' branches of the food processing industry.

In other respects, backward food processors show many features in common with backward operators in other sectors. Processing is often done on a parttime basis, there is no formal organization of operations (e.g. no formal accounting and marketing arrangements), the techniques used make processing arduous and unpleasant, and operations are generally undercapitalized so that the use of hard cash is avoided as much as possible. From this it is evident that, while backward processors have a very favorable resource orientation, they do not exert a significant stimulus on primary production of agricultural products.

Typically, the major part of the raw material needs of the advanced branches of the industry is met by imports. For example, a survey(5) of 38 nongovernmental food and beverage manufacturers found that 64 percent of the 77 products they produced derived more than half of their raw material inputs from extra-regional sources (excluding energy, which is also import-based for the most part). Yet, only 22 percent of the 77 products required a substantial input of specifically foreign raw materials using the installed technology. It is fair to say that as presently organized the industry functions as part of the import and distributive sector rather than as a truly indigenous development of industrial activity based on available resources. Indeed, traditional importers see food processors as competitors (not as suppliers) because, in effect, the establishment of a local food processing industry simply transfers earnings − such as importer's commissions and wholesale mark-ups − from one entrepreneur to another, with hardly any increase in the level of local benefits.

Monopoly appears to be the norm for the industry. Entrepreneurs are fearful of competition and are unwilling to divide up markets. Thus, we find very often a single enterprise controls the national market for a product, and plant size matches (or more correctly is geared to) market size. Moreover, in those branches of the industry where monopoly and dualism are not fully developed, there are strong mechanisms and processes in train which are forcing the disappearance of intermediate-level operators. For example, in the baking industry, the larger operators are taking steps to differentiate their products from those of backward and intermediate-level operators by a process of mechanization linked with vigorous advertising. They promote product character-istics such as purity (absence of human handling), high quality (fine textured and plastic), etc., which require a level of mechanization intermediate-level operators cannot afford. The proportion of sales going to the latter operators is diminishing, and those who attempt to mechanize as a countermeasure are going bankrupt.

Other means are in operation to maintain monopolies by discour-aging entry of new processors. For example, one finds that many firms who have already established dominant positions in their national markets, using their own product formulations, nevertheless enter into licensing arrangements with foreign firms to produce substitutable brand-named products for the same local market. Such steps are clearly designed to pre-empt opportunities for other entrepreneurs to enter their markets as strong competitors. In almost every instance, the use of foreign brand names operates against the interests of national development, as the quality control or technical support provided under such arrangements discourages indigenization of the raw material and technical base of the industry.

The demand for technology is directed almost entirely towards extra-regional sources. Foreign dominance of industry operation is initiated and maintained by the very limited linkages established between technology users and national engineering firms (even when greater linkages are possible), by the absence of significant adaptations and innovations in the direction of greater use of native resources, and by the absence of measures to ensure that external economies associated with private technology acquisition are tapped to the fullest. Furthermore, the involvement of foreign consultants tends to insulate technology users from benefits of competition between foreign technol-ogy suppliers, quite apart from reinforcing dependent relationships with industrial economies. Evidence that foreign consultants show preference for certain suppliers implies that the preferred suppliers can act as virtual monopolists. Government policy measures, such as duty-free importation of machinery, which raise barriers to the design and fabrication of machinery by indigenous engineering firms, exacerbate the problems of developing an indigenous scientific and technological capability and promoting the spread of modernization. Thus, foreign technology is being supplied to the industry on terms which are most advantageous to the foreign technology supplier, and in ways which make little contribution to the erosion of foreign dependence and the unfavorable international distribution of benefits arising from the activities in the industry.

A number of minor innovations and adaptations do occur constantly within the industry. These include product variations; modifications to machinery to speed up operations, to conserve energy, or to enable the performance of additional or alternative tasks; and modifications of processing methods to alter the characteristics of the finished product. There is also copying and fabrication of intermediate level equipment and tools, and a few instances of copying of simple components of modern production plants (such as drying ovens). Typically, such innovations, adaptations, and copying are the results of efforts by individuals within productive enterprises using essentially empirical approaches. The services provided by national engineering firms are, almost exclusively, repair and maintenance of equipment, and plant installation. With the exception of the most backward operators, who show no tendency to innovate, the key development processes of innovation, adaption, and copying which are emerging spontaneously at various levels of technological advancement (particularly the inter-mediate level) receive no formal support, and in many respects are suppressed by the way the industry is integrated with the rest of the economy.

The small size of national and even regional technology markets is an important structural factor affecting the extent to which indigenous engineering firms can become involved in meeting the needs of the industry under existing patterns of development. The complete assimila-tion of a significant proportion of the production equipment and techniques currently used by the industry will require a degree of specialization in engineering skills and in equipment production which cannot be supported by the industry. Hence, substantial progress in the assimilation and control of technology, as the industry grows, will depend on the ability to identify or develop production technology which can be copied, fabricated, and adapted without investment in highly specialized and costly equipment and skills.

Difficulties Associated with the Backward Linkage to Agriculture

The problems of producing low-cost processed foods based on indigenous materials is usually considered to be located mainly in primary agriculture; that is, it is seen in terms of the low productivity of farmers which result in the cost of indigenous raw material inputs being high in relation to the price the finished product can fetch on the market. For example, in 1975, payments for whole yams accounted for 50 percent of the production cost of 'instant yam,' a product produced in Barbados from a traditional food crop. Poor quality, or variability in quality, and the unreliability of supplies from indigenous agriculture are additional limiting factors in the development of linkages.

The agricultural export sector has been the greatest beneficiary from R&D in agriculture and, since small farmers account for a major proportion of food crops grown for local consumption, these problems are attributable, in part, to a failure to provide appropriate R&D support for small farmers and for the development of indigenous food crops. However, it is fair to say that, using existing technology, yields per acre can be increased several-fold in many cases, but the existence of perennial gluts on the fresh crop market discourages farmers from trying to do so. The development of linkages between indigenous primary producers of food crops and the food processing industry is caught in a vicious circle. Without substantial additional investment and effort, the low yields in primary agriculture (which lead to high costs), plus the yearly variations in quantity and quality of produce, will combine to discourage the use of local agricultural products by the processors. Without the development of processing facilities to receive the freshmarket surplus, there will be no attempt made by farmers either to increase production or to produce special varieties in order to provide the adequate raw material base which encourages investment in processing. Not unexpectedly, one finds that farmers and processors blame each other for the impasse.

The usual form of government intervention, which is to establish and subsidize a central processing plant operated along conventional lines, fails miserably in its social and economic objectives. Such linkages as are created between central processing plants and primary producers tend to involve only the modern, larger farms and plantations, which then receive handsome subsidies at the taxpayer's and/or consumer's expense; while the groups in the backward sector, badly in need of stable incomes, are denied access to such opportunities by the way the industry becomes organized. For example, in Barbados, a milk processing plant, established with government participation, subsidizes (by operating at a loss on the fresh milk market) about 25 relatively large dairy farmers, who supply it with milk from about 1,000 cows. But under the existing organization of the milk processing industry the plant finds it impracticable to utilize the output from about 2,000 other cows, owned by small farmers, because it is either 'uneconomic' to collect their output, or the small farmers cannot afford to install handling and storage equipment to meet the processing plant's requirements.

Development Objectives for the Food Processing Industry

Development planning for the food processing industry should be concerned with increasing both the nutritional quality and the total supply of food available to the population of these states, while, at the same time, ensuring there is rapid progress along the two lines of development identified earlier. Since the vast majority of the ECCM states and, to a lesser extent, Barbados, remain essentially agricultural economies, a crucial objective will be to ensure that agriculture develops as food processing industries develop. Finally, assuming there is a commitment to development within a regional framework, the strengthening of the economic integration movement should be an important development objective. On this basis, one may identify the following first-round targets for future developments in food processing activity.

Increasing (regional) self-reliance in food production

Self-reliance in this context must be seen as an important strategy for maximizing the benefits gained from the use of national resources, as well as for safeguarding national and regional self-determination. Efforts must be concentrated on satisfying basic nutritional needs of the population using indigenous raw material, capital, and technology, as far as possible, so that there is eventual emergence of national control over the direction of development of the subsector. The achievement of these objectives will require that strategies be found which will lead to a progressive increase in the scope for indigenous involvement in meeting the industry's needs. In this regard, it will be advantageous to approach this goal from two directions: by stimulating and coordinating learning processes which will lead to an accumulation of skills and knowledge needed by the industry, as well as by simplifying the processing requirements, plant designs, and techniques needed by the industry.

Harmonization of regional industrial development to promote equity in economic exchange between member states and within member states of CARICOM

What is needed here is a rational basis for industrial allocation between member states and within member states of the integration movement, so that all states and all social groups within these states derive economic benefits which they consider adequate and fair in relation to any costs of participation. In essence, a mechanism for engendering strong political and popular commitment to regional cooperation in production must be quickly established and all obstacles to cooperation dismantled. This task is all the more difficult because of great similarities in the production potentials of the various states.

Maximization of the flow of benefits to the poor from activities in the food processing industry

This objective requires that strategies be devised for promoting the greatest flow of benefits to low-income households from all food processing activities and activities linked with the food processing industry. Specifically, means of achieving greater efficiency in con- sumption, as well as production, by the poor will have to be found, so that their purchasing power and assets are used to maximum advantage − economically, socially, and politically.   Access of low-income householders to opportunities for economic and social advancement will also have to be facilitated, so that there can be a steady rise in their asset-forming and political power.

Stimulation of growth and development in agriculture

The achievement of this objective requires that production methods, and a production structure, must be developed in the food processing industry which will make the linkage problems originating with primary agriculture less restrictive than they have been in the past. In particular, modes of operation must be chosen which will maximize the involvement of small farmers, rather than exclude them from the industry, both because of their traditional role as the principal local supplers of food and because over 90 percent of farms are small.

Relevance of Intermediate Technology to Development Objectives

All three strategies proposed earlier can be shown to be highly applicable to the development problems and needs of the food processing industry in the ECCM and Barbados.

It is possible to understand the importance of the optimal-spread strategy to the development of the subsector by applying it to the ways in which consumer needs are to be met. The achievement of adequately remunerative employment will, clearly, be strongly dependent on the cash income needed to meet one's needs for survival, such as food. A poor choice of food products can worsen the position of low-income groups, and an appropriate choice can strengthen their position. Similarly, since product choices are ultimately reflected in choice of technique, which in turn determines opportunities for local skill development and the demand for labor, the set of products produced in the subsector will strongly determine its capacity for internally-driven development and for labor absorption.

Thus, recognizing that processing increases food cost but tends to reduce, rather than enhance, the nutritional value of the basic materials, products selected for manufacture should reduce the level of processing required, compared to the imports displaced. Reduction in processing would not only keep food costs down but should also make such products amenable to production using simple techniques. Another important implication of the optimal-spread strategy for the food processing industry is that governments should ensure that the market for fresh food is fully exploited, or very nearly so, before it can be used as a raw material in processing. In this way processing would not attract material away from fresh consumption, which is both cheaper and more beneficial to the consumer.

In the context of an industry where stringent health standards cannot be compromised, the subcontracting strategy can be applied as follows. First, production techniques which do not depend critically on process or quality control and do not pose food safety problems can be reserved for intermediate level, small-scale operators. Second, processing of crops susceptible to damage in transportation, or to rapid nutritional loss in storage and transport, could be allocated to a network of small processors who could produce a semiprocessed product which obviates these problems, while reserving final touch and quality control measures for a modern centralized plant.

The pilot plant strategy could be used with advantage to overcome the linkage problem with primary agriculture. Until processing facilities are established, the food crops available for processing will remain small, varied, seasonal, and dispersed. There is clearly a need for small processing plants which are flexible enough to take advantage of small surpluses from the fresh produce market which become available at different times of the year and, eventually, stimulate increased production for the processing industry by providing additional secure markets for farmers.

NOTES

(1) H.R. Brewster, "Export, Employment and wages in Trinidad and Tobago," CSO Research Paper No. 5 (1968).

(2) E.F. Schumacher, First Sir Winston Scott Memorial Lecture, Barbados (1976).

(3) See for example Secondary Agro-based industries Sector Study, Caribbean Technology Policy Studies Project.

(4) The information used here is drawn entirely from J.W. Dellimore and J.A. Whitehad, "Secondary Agro-based Industries Sector Study," Caribbean Science and Technology Policy Studies Project.

(5) This survey was conducted as part of the Caribbean Science and Technology Policy Studies Project, Secondary Agro-based Industries Study.

8
Commercialization of Technology and Dependent Underdevelopment— Some Current Issues and Future Policies
Maurice A. Odle

This chapter seeks to examine the process of commercialization of technology from the point of view of the underdeveloped countries of the Third World. As such, it makes no pretense at trying to represent the "interests" of the technological suppliers who derive most of the benefits from the unequal system and, in any case, can take good care of themselves. Nor does it attempt to treat with the trade in technology among the developed countries.

Cognizance is taken of the fact that a United Nations Conference on Science and Technology for Development is scheduled to be held in August 1979 in Vienna. We are also aware that manufacturing technology, though very important, is only one of the variables in a social matrix that needs to be transformed before a new international economic order can be achieved. Just as important is the need for new systems of commodity arrangements and international finance (at the moment mainly for infrastructure) which can assist in promoting effective national and regional industrial programing. The old systems are all interrelated in one composite unequal economic order.(1) For example, low priced primary commodities from the periphery are in unequal exchange for high priced industrial goods from the metropolitan centers. The resulting finance gap requires seeking aid from the very centers which, in turn, are usually tied to an inappropriate foreign technology.

In Section 1, we shall review the progressing stages of development of the commercialization of technology, as a guide to prescriptive action. Section 2 discusses the homogeneity of supplier thinking and objectives within the international capitalist system, and the consequent strategic limitations in extracting greater benefits within the present system. Section 3 examines the heterogeneous nature of the Third World economies – Latin American, Caribbean, Asian, and African – their differential treatment, and the theoretical and policy implications. Section 4 takes a critical look at certian past, present, and potential recommendations for international, regional, and national

98

action; and Section 5 attempts a summary and conclusion.

SECTION 1: ANALYTICAL STAGES IN THE APPROACH TO COMMERCIALIZATION

From the point of view of the developing countries, the problem of the commercialization of technology is historically linked with the movement of foreign capital, in the same way that the MNC phenomenon was associated with foreign investment (in mining and plantation agriculture) from the very inception of the colonial era. Although technology flows today are mostly one way, in the precolonial/ preindustrial period a considerable amount of technology actually moved in the reverse direction. Present-day developed countries are indebted to the underdeveloped for many agricultural, industrial, military, and navigational developments. During the colonial period the technology was simply borrowed rather than commercialized. However, the approach to the issue of commercialization varied considerably over the years, partly because of the nature of foreign capital itself, and partly because of the changing consciousness and perceptions of the analysts.

Transition from the Implicit to the Explicit

During the early colonial period, technology was given implicit treatment and assumed to be embodied in the foreign capital. This was possible because the technology was of a proprietary nature and imported along with a foreign subsidiary. There was no need on the part of the owners of capital to make the technology explicit, in an accounting sense, because there were no restrictions on the rate of return that could be earned or on the level of profits that could be repatriated. The consequences of any embodied or, even more impor-tantly, disembodied technological change that occurred were, therefore, subsumed under general profits.

The change from an implicit to an explicit treatment of technology began to occur in the interwar period and accelerated in the 1940s when certain underdeveloped countries started to place curbs on the tax allowances and profits of the foreign subsidiaries, and to limit the amount that could be repatriated. One or two MNCs began to introduce arms' length agreements instead of the traditional tacit or gentlemen's agreements. The possibility of nationalization accelerated this process even further, as did the incidence of registration of patents, both designed to ensure that the technology was not actually transferred. For example, there was a spate of registration of mining patents after Allende nationalized the copper companies in Chile. As a result, most underdeveloped countries ended up signing technology contracts after takeovers. Nevertheless, the greatest fillip to the explicit commerciali-zation of technology occurred with the embracing of the policy of import substitution by most underdeveloped countries after World War

II. Tariffs were placed on many manufactured goods in order to stimulate local production and reduce economic reliance on narrowly-based, primary export activity. In order to maintain access to the local market, tariff-hopping MNCs were eager to set up subsidiaries, to enter into joint agreements, or to sign contracts with local producers. In most of these cases, there existed arms' length contracts. In addition to high fees and heavy overinvoicing on tied inputs in all three cases, the subsidiary, operating in a protected market, was able to earn large profits (a double reward). The local technology recipient, for his part, instead of involving himself in the costly and uncertain process of indigenous research, preferred the well-tested package of know-how, equipment, organization and trouble-shooting that the supplier had to offer; for him, private costs, rather than social costs, were the important factor. It is this later period, involving the proliferation of contracts, that is commonly associated with the concept of the commercialization of technology.

The academic approach during this early stage was, as expected, indirect rather than direct. This was due to at least three factors. First, the transition from implicit to explicit technology was a very new and recent phenomenon to which everyone was not alive.(2) Government agencies thought that by keeping the contractual information secret they were safeguarding the national interests; this failure to exchange information meant that in the process they were even more exploited.(3) Second, many writers had rather casually assumed that technology was actually transferred by the MNC, even though they recognized that a large part was not only inappropriate (e.g. excessively capital intensive and nonlabor absorbing), and also somewhat dysfunctional (in the sense that technology was used to produce minerals and plantation crops that were entirely exported, or luxury goods, and was not catering to the basic needs of the mass of population). They, therefore, tended to concentrate on issues pertaining to ownership, and to neglect the problem of technological control. Third, some writers were still preoccupied with the issue of fair-pricing ("terms" of trade) of the object or product of the technology.

Reducing Costs Approach

This approach is most closely connected with a series of case studies, commissioned by the Intergovernmental Groups on the Transfer of Technology, and undertaken by(4), or on behalf of(5), the UNCTAD Secretariat in the early 1970s and which were published in 1974.(6) These studies, although written from the point of view of the recipient, are nearly all lacking either a developmental framework, or a theory of the relations between the center and the periphery. In an apparent attempt to maintain international neutrality and to avoid ascribing blame (for the lack of development) to either suppliers or recipients, many studies ended up devoid of all social value. It is tacitly assumed that a substantial part of the technology is actually being transferred (under an apparently basically equitable system), and so no effort is made to identify the parties who derive the greatest benefits and gain from the international trade in technology.

The approach of the authors, therefore, is to concentrate on merely describing the commercialization process, and at certain points to suggest the need for a reduction in costs. Admittedly, the authors are concerned not only with direct costs (that relate to various types of payment, including transfer pricing), but also with indirect costs associated with the myriad restrictive practices. Nevertheless, the studies remain in very much a positive, rather than a normative, vein. There is also little attempt to assess the effect on such government objectives as faster growth, full employment, balance of payments equilibrium, stable prices, equitable distribution of income, and the impact on the long-run developmental prospects.

Primarily on the basis of the commissioned studies, UNCTAD devised a transfer of technology code in 1975 without, as expected, any particular philosophical underpinning.(7)

Dependent Underdevelopment Approach

At the same time that the United Nations studies were being commissioned, a more radical and independent body of opinion was being developed. The dependent underdevelopment approach stemmed from two streams of thought. First, there were the authors from the Latin American(8) and Sussex(9) schools who, having analyzed data not previously available, came to the conclusion that technology was not being transferred at all (nor was ever intended to be) owing to the various restrictive practices.(10) Not even the mere techniques were being transferred since there were restrictions on their use at the end of the life of the agreement.(11) The technology is, therefore, merely leased or rented. Second, this view of nontransference has been accepted by some members of the same schools, plus other radical scholars, and integrated into the general theory of dependent underdevelopment.(12) The scholars are concerned to show that the present-day process of commercialization of technology is one of the major factors contributing to dependence and underdevelopment, with increasing marginalization of the periphery in world production (and science), and widening disparities in income both within each periphery territory, and between the periphery and the center. According to the clear exposition of one author,

> the transfer of technology from industrialized to underdeveloped countries is tending to increase technological dependency. Foreign technology tends to be highly differentiated, capital-intensive, large-scale, and associated with the production of luxury consumer goods ... When transplanted into a socioeconomic context where there is large unemployment, where such labor skills as exist tend to be artisanal, where markets are traditionally small and where the needs of the mass of the population are for the basic necessities of life (food, shelter,

clothing, health) the distortions become obvious.(13)

Other writers have contributed to the development of the theoretical relationship between technology, dependence, and underdevelopment by making the distinction between apparent industrialization and real industrialization.(14) Still others have elevated the basic needs philosophy and given it even sharper focus. Proponents of this theory assert that technology is only relevant and worth importing to the extent that it serves the basic needs of the people. They, therefore, insist that countries should produce what they consume and consume what they produce. In this scheme of convergence of resource use and demand, exporting, for example, would be a mere extension of producing for the home market. According to one author, then, an important measure of economic backwardness is

> on the one hand, the lack of an organic link, rooted in indigenous science and technology, between the pattern and growth of domestic resource use and the pattern and growth of domestic demand, and on the other hand, the divergence between domestic demand and the broad mass of the population.(15)

This is the only way of resolving the dilemma and dialectical situation posed by a technology which, while recognized as being essential for transformation, when applied to certain contexts only succeeds in helping to retard development.(16) The stage reached by recent theorizing is reflected in the view that "the technological dependence of the Third World is not only a reflection of its economic dependence, but is also an active agent in perpetuating this domination."(17)

It would appear that the United Nations has now come around to this point of view. According to a recent report by the UNCTAD Secretariat,

> The inappropriateness of many of the rich country products introduced and promoted in the domestic markets of poor countries derives from the fact that they embody technological characteristics that are either unnecessary, undesired or too costly to meet the basic needs of nutrition, health, clothing and shelter. Labor-intensive methods of production are sometimes excluded if these modern products of high quality are to be manufactured . . . In sum, the efforts needed to respond to the basic wants of the great majority of the Third World's population are beyond the field of interest of the transnational corporations . . . Dependence is built into this industrialization process. Moreover, the technological dependence of developing countries may be self perpetuating.(18)

SECTION 2: HOMOGENEITY OF SUPPLIERS

Given recent developments, the idea seems to have won out that what the Third World requires is a basic needs strategy for development. To effect this, it appears that two major tactical measures are being recommended – unpackaging and alternative sourcing. While these policies merit implementation, we feel that sufficient recognition has not been given in the literature to the homogeneous nature of the supplier system.

Basic Needs

There are at least two political and sociotechnical problems associated with the basic needs strategy.(19) First, such a fundamental change in the approach to development would depend on a basic ideological shift in the typical Third World country. Current political hierarchies embrace new philosophies only very reluctantly, because they are entangled in, and benefit from, the existing system in very many ways. Local capitalists, of course, have a vested interest in maintaining the elitist status quo, and command greater access to the seat of power than those who might benefit from an alternative set-up, in which income might be more evenly distributed. The suppliers, for their part, would probably attempt to wield influence either directly, or through their governments and international funding institutions. Obviously the major suppliers (especially those whose governments are imperialist powers) would try to defend in the highest forums the interests of the weaker suppliers, believing that any breach in the system would threaten the interests of all.

Second, this strategy emphasizes basic goods but not basic technology, even though the former would in many cases automatically lead to the latter. For example, though capable of increasing the supply of food, the technology employed in the green revolution was mostly suited to large farmers who had the funds to make extensive investments in drainage and fertilizers. (This has important distribution of income implications.) Similarly, clothing and shelter can be provided by either capital-intensive or labor-intensive technology, although admittedly a choice exists, whereas it seldom obtained for technology-specific and highly-differentiated consumer durables.(20) The problem remains that all suppliers are interested in selling goods (whether consumer, intermediate, or capital) embodying the latest technology, rather than intermediate technology.

Unpackaging

Once specific economic and social goals (e.g. basic needs) have been decided on, certain technology policy instruments, such as unpackaging (and alternative sourcing) can be applied. Unpackaging (or "opening up of the black box") refers to the disaggregation of the technological input into its core components (such as processing and basic engineering), and its peripheral components (civil, foundation, construction, and electrical engineering, design engineering, technical assistance in plant layout, and choice of equipment). This requires careful appraisal of the ways in which technology is imported (and the related tied inputs) so as to bring about

(a) the rationalization and better negotiation of these imports by breaking the ties that generate monopolistic profits; (b) the creation of demand for local activities that could be developed efficiently; and (c) the assimilation of imported technology and its complementarity with innovative efforts within the importing country.(21)

The unpackaging policy, therefore, laudably seeks to maximize the actual and potential local technological capability. However, the policy may have only limited effect for both historical and market structure reasons. The behavior of the technology suppliers (who are now alert to the unpackaging strategy and capable of countering accordingly) is much more standardized, with the market dominated by the multinational phenomenon, than it was during the period of industrialization of Japan, the classic, and perhaps only, unpackaging success story.(22) Today it is thought that, even though it may be possible to disaggregate for certain industries.

there are also important sectors – and subsectors – where 'packaged' technology systems supplied by machinery firms directly dominate, and where unpackaging by the customer (assuming that he wishes to do so in the first place) is difficult. This particular finding is probably rather unsurprising to anyone who has experience of (for example) the role of contractors in machinery markets. But the fact is that it has not been recognized in a systematic way by policy makers concerned with 'transfer of technology' nor by academic researchers studying the process.(23)

This suggests that suppliers are able to use their superior bargaining power, in a frequently oligopolistic market, to insist on the sale of a joint, rather than a disaggregated or singular, product. There is homogeneity among capitalist suppliers as far as motivation and aims are concerned, and it is only differences in recipient bargaining power, information-gathering, and capabilities that produce different results. The little available evidence seems to suggest the need for international governmental intervention in this imperfectly competitive situation and/or a different social formation at the international level.

Alternative Sources

Unpackaging is really part of the search for alternative technologies which maximize the use of local factors of production and inputs. A complementary activity is the search for alternative sources so as to increase the capacity for bargaining and selectivity. This policy instrument is very much dependent on gathering as much relevant information as possible.

As a policy instrument, alternative sourcing also has limitations. First, technology suppliers are all profit-maximizers and so differences in behavior among them would tend to be marginal. Second, suppliers themselves are probably aware of the conditions others offer. Third, there is probably a certain amount of collusion and informal cartelization on the part of suppliers. There may even be a tacit agreement for the staking out of certain parts of the Third World as spheres of influence (or leadership) of particular suppliers/countries.

The experience of the English-speaking Caribbean for packaged technology certainly seems to indicate that there are no discernible differences among suppliers. A recent study(24) found that there were no significant differences between developed capitalist suppliers (or between the most powerful and least powerful countries in each group), whether they came from traditional sources (Britain, United States, and Canada) or nontraditional sources (Western Europe and Japan). Suppliers in technologically intermediate, but yet underdeveloped, economies (e.g. India) also did not impose less onerous conditions. This is not unexpected since capitalists, wherever they may be, are all motivated to maximize profits. (Conversely, the contracts with the socialist countries were deliberately designed actually to transfer technology.) The differences between the consumer, intermediate, and capital goods categories, and between sectors on the basis of technological endowment (e.g. traditional vs. modern and standard technology vs. frontier technology), were also unexpectedly small (except in the case of petroleum), and it did not seem to matter whether recipients were local capitalists, a local/foreign consortium, or government.

The implications are that the international search for technology of an alternative and disaggregated type, to suit the requirements of unpackaging, is likely to bear only limited fruit. Unless international action can somehow change the rules of the game in the trade in technology, underdeveloped countries may find it necessary to disengage from trade with capitalist countries and establish links with probably the only really alternative source − developed socialist countries.(25) This change in trading partners is feasible in the context of a basic needs strategy, since any technological advantage of capitalist countries over socialist states lies primarily in the area of highly differentiated consumer durable products; the developed socialist countries may, therefore, also be the source for the only genuine alternative technology.

SECTION 3: NON-HOMOGENEITY OF RECIPIENTS

It is now becoming recognized that at least one thing that local private recipients in various underdeveloped countries have in common is a lack of concern for the indirect or social costs associated with the importation of technology. For example, it matters little to them whether inputs are acquired locally or abroad (ignoring the problem of overinvoicing) provided that profits are maximized. (This shows the extent to which there is need for national and international government intervention.) However, until very recently, little recognition was given in the transfer of technology literature to the differences between recipient countries, and the contexts in which recipient enterprises operate.

Stages of Underdevelopment

Despite their general underdeveloped status, there are wide differences between Latin America, Asia, and Africa in terms of per capita income and other worthwhile indexes.(26) Of course, there are also important differences within each of these continental groups; for example, in Latin America there are significant differences between the larger territories (Mexico, Argentina, and Brazil) and the rest. For classification purposes, the underdeveloped world can, therefore, be divided into the intermediate economies (e.g. the larger Latin American countries and India), the modally underdeveloped (the bulk of the Third World), and the least developed (comprising about 25 states according to World Bank ratings). This classification has significance for determining the ability to bargain for technology deemed most relevant and least costly by the various countries. For example, the least developed countries might experience grave difficulties in bargaining for technology because they lack a competitive local technology. For the same reason, they might be expected to make the least progress with an unpackaging strategy.

The Size Factor

Size is also an important factor in relation to importation of technology. Although there is strong correlation between market size and stage of underdevelopment, there are some fairly large economies (e.g. Ethiopia, particularly from a population point of view) which are still classified as least developed. Large economic size makes possible the production of certain types of products (e.g. capital goods), although the advantage of market size is not as great with respect to highly differentiated, elitist, short plant-run type consumer durables (especially if the latter are merely assembled). Size is also important in the bargaining process, as it allows the recipient to negotiate more effectively. A supplier is much more likely to concede if he runs the danger of losing, or foregoing, a large market and its attendant absolute level of profits. In this regard, the collective weight of a common market (comprising recipients) can be important. This point is recognized by the Acuerdo de Cartagena in its reference to "the joint acquisition of similar technological inputs that are needed by two or more countries."(27)

Some Empirical Evidence

There are very few studies comparing the terms and conditions under which different recipient countries acquire technology. (There is also a lack of comparison of the terms and conditions under which types of recipient enterprises, in a particular country and between countries, receive their technology.) However, there are available raw data on individual countries which could be useful for comparative purposes.

The intermediate and large Latin American economies (Argentina, Brazil, and Mexico) and the Andean Pact countries (Bolivia, Chile, Colombia, Ecuador, Peru, and Venezuela) have, within recent years, generally imposed limitations on royalties (usually 5 percent, with it being taxable) and trademark payments (normally 1 percent), payment between related (MNC – subsidiary) parties, the duration of payments, the length of contracts (in most cases, five years), the capitalization of technology and various restrictive practices.(28) India also limited the contract duration to five years, and the royalty rate to 5 percent, and levied a tax on payments of royalties and technical fees, based on the difference between the fees and the costs incurred by the enterprise in earning them (50 percent on agreements made since 1961 and 70 percent on agreements made before that date).(29) There may be one or two other underdeveloped countries (e.g. Malaysia who are not doing too badly.(30)

On the other hand, most of the other underdeveloped countries have not felt strong enough (or have been too unaware of the degree of exploitation) to impose any limitations. For example, the Caribbean Technology Policy Studies Project recently found that of the available technology contracts for Guyana and Trinidad (two supposedly modally underdeveloped economies) 11 out of 30 and 29 out of 49, respectively, were longer than five years in duration.(31) And a similarly recent breakdown of the Jamaican contracts into licensing (of know-how) and management types revealed that 33 out of 71 and 9 out of 22, respectively, had a life in excess of five years.(32) With respect to payments, an equally large proportion of the contracts relating to licensed technology exceeded a 5 percent of sales figure; for example, in Jamaica, this was the situation with respect to 28 of the 46 contracts, with some cases recorded of charges of 20 percent of sales and over. In addition, these percentage of sales charges were sometimes combined with a lump-sum imposition or some other type of payment.(33) Similarly, in very many of the contracts, an extremely large number of restrictive practices were present, including those that were of critical significance to not only the   recipient enterprise but also the recipient country. The situation is probably even more pronounced for the least developed economies. In the case of Ethiopia, for example,

> the costs involved in the transfer process amounted to some 2.8 percent of GDP, to over one-third of export proceeds and to a little more than half the net value added in the modern manufacturing sector. Comparable figures for developing countries as a group have been estimated to be – again with many

qualifications – under 1 percent of their combined GDP and around 4 to 5 percent of their proceeds. The sharp contrast between Ethiopia – the largest of the least developed countries – and the developing countries as a group serves to underline the severity of the burden of costs of the transfer process on least developed countries.(34)

Although the above evidence is not altogether convincing, the apparent implications are that any policy recommendations need to recognize the even more vulnerable and special position of the really underdeveloped (and frequently small) economies of the Third World. UNCTAD has only recently begun to give explicit recogniton to these differences. In one of its most recent position papers, it recommended a "program of international cooperation and assistance" so as to increase technological capability and bargaining power. "There would be immedi- ate savings to developing countries in that the costs of transferring technology would be reduced; such savings have already been attained to a significant degree by, for example, Argentina, Colombia, and Mexico."(35) The paper also recognizes the more favorable position of "the more industrialized among the developing countries" as far as remedial contractual and institutional arrangements are concerned. "The process has been particularly marked in Latin America, where some of the Andean Pact countries, Argentina, Brazil, and Mexico, have initiated measures affecting importation of technology; and correspond- ing measures have been introduced in some other countries, including India."(36) In referring to the fact that only 2 percent of total R&D in the world is carried out in developing countries, the paper makes the further distinction that "much of the 2 percent is inevitably carried out in the more advanced of the developing countries"(37) partly because "the education system and the level of education in Latin America, most of North Africa and India is more advanced than in the rest of the developing countries."(38) This differentiation between developing countries needs to be much more clearly reflected in policy proposals for a new technological order. In its discussion of appropriate national technology centers, UNCTAD also made a distinction between seven types of economic situations: (a) Large country, open economy, high on the developing scale; (b) Medium-sized country, well up on the developing scale and rural-oriented; (c) Medium-sized country, rising on the development scale, large state sector and heavy-industry oriented; (d) Medium-sized country, middle of the development scale, open economy, and switching from import-substitution to export-oriented industrialization; (e) Mining economy, fairly high GDP, but low on the development scale and devoid of transfer of technology institutions; (f) Poor country, planned economy, with emphasis on rural sector and redistribution of incomes; and (g) Least developed predominately agricultural, and totally lacking in technological policies, institutions, and infrastructure. In the past, the more clearly articulated views of the intermediate economies seem to have dominated the thinking of the international policy makers and, consequently, proposals for reform. Of course, the intransigence of the developed countries was also an important factor. Ironically, although these countries were the first to become conscious of the unequal nature of the trade in technology, and initiated thinking and reform, their proposals never went far enough to take adequately into account the weaker position of the bulk of the developing countries.

SECTION 4: INTERNATIONAL, REGIONAL, AND NATIONAL ACTION

There is great imperfection in the international market for technology. First, all suppliers are not known to all buyers and this market segmentation is compounded by the fact that terms negotiated only by a very few buyers are not known to other buyers. Technology recipients hide information from other recipients (whom they imagine are their competitors), and are protected by archaic disclosure laws by their governments who believe that such secrecy is in the national interest. Recipients are, therefore, ignorant as to what is, or should be, the norm. The net result is that price discrimination is rampant. Second, suppliers of technology have a monopoly of knowledge, and recipients often possess very little information about the nature of the commodity being purchased. This allows the supplier to sell the technology in a packaged form, and to include a number of inputs that the recipient either does not really need or could supply locally. The recipient, on his part, only reinforces the noncompetitive situation by his desire for a ready-made, well-proven, and trouble-free product.(39) Third, the structure of the market gives rise to an inability to calculate and to make determinations, a situation frequently expressed through gross multiple pricing for the same service. For example, a piece of machinery embodying a certain technology may have a stiff ex factory price, the supplier may indulge in transfer pricing (overinvoicing), high profits may be earned, and a large royalty imposed, all for the same technology.(40) Such high degrees of market imperfection require intervention at all levels.

International Action

Many types of international action have been proposed over the years but there has been little implementation to date. For example, an international code of conduct for the transfer of technology was bruited about since the beginning of the 1970s, but the first real attempt at formulation only occurred in 1974 at the Pugwash Conference.(41) In 1975 an UNCTAD commissioned International Group of Experts expanded on the Pugwash code "taking into consideration particularly the needs of developing countries and the legitimate (sic) interests of technology suppliers and technology recipients."(42) It would appear that the needs of developing countries are not consistent with the interests of technology suppliers and the objections of the supplier countries prevented agreement being reached. In the 1976 UNCTAD IV meeting, there were again "divergent positions," with the developed countries insisting that "a voluntary code of conduct would best serve the transfer of technology," whereas the underdeveloped countries "firmly believed that a multilateral legally-binding instrument was the only way of regulating effectively the transfer of technology to developing countries."(43) It is clear that developed countries want to retain the benefits derived from nontransference of technology.

But even when introduced, the suppliers would seek to avoid and evade the strictures of the code. For example, gentlemen's agreements between parent companies and subsidiaries are open to abuse. Second, technology recipients in the private sector, acting out their international comprador bourgeoisie role, may be willing partners to tacit agreements with the suppliers. Third, suppliers might devise counter strategies to prevent technology from being really transferred.(44) We feel that future technology contracts will have the barest minimum of information unless the state stipulates otherwise. Nevertheless, a code is of vital importance to the less developed of the Third World countries. Whereas the intermediate economies have been able unilaterally to introduce national codes, the other underdeveloped countries feel they lack the bargaining power to do likewise.

Of course, a code is a necessary, but not sufficient, condition for the technological development of the Third World. Patent laws, for example, need to be fundamentally revised so as not to be against the interests of the developing countries. Laws should establish automatic compulsory license of right, impose the burden of proof of exploitation of patents on patentees (with administrative rather than costly and delaying judicial procedures), and prevent the imposition of patents on imports.(45) Again, such legislation is of particular importance to the less developed countries of the Third World,(46) even though at their present stage of consumer-goods-oriented industrialization (which highlights product differentiation), trademarks are probably more important than patents, in contrast to the situation that would obtain if the intermediate and capital goods sectors were more greatly stressed.(47) Again, some intermediate economies have been able to ignore the system (e.g. Brazil), or to prevent patenting in certain vital sectors (e.g. India), or generally to introduce the necessary national legislation proscribing the worst aspects of the Paris Convention. As expected, vested interests of the supplier countries have so far prevented the required revision, and the matter is still under discussion and review.(48)

There are other supporting international measures. One is setting up of an Information Center where contracts and experiences recorded from national sources can be registered. Such evaluated information on alternative technologies and sources of supply, and terms and conditions, could be of inestimable value to all importers of technology. We feel the information system could be of most use to the less developed of the Third World countries, who are less aware and less organized; in particular it could be a forerunner to the introduction of a most favored nation treatment system, a device which needs to be stressed more at UNCTAD forums. The exchange of information could also assist the less developed of the Third World countries in identifying cases of transfer pricing, since the manpower and organization effort involved in effectively detecting overinvoicing and underinvoicing may be beyond the resources of any one of these countries. Data on the Caribbean are not easy to come by. But in a published advertisement in the Trinidad Guardian (Dec. 3, 1977), the Oilfield Workers Trade Union revealed that Federation Chemicals in 1976 sold to its parent company, W.R. Grace of the USA, urea at an average of TT$186 per ton. But at the same time it was selling urea to Guyana at an average price of TT$330 per ton. (Similar overinvoicing exists with respect to ammonia, Federation Chemicals' principal output, and these experiences may be indicative of the situation in the Caribbean as a whole.)

The reaction of the developed countries at UNCTAD IV to the proposal for an Information Center was predictable. They stipulated that the appropriate exchange of information on technological alternatives available to developing countries "must be consistent with contractual agreements and, (my emphasis) where relevant, must respect the confidentiality of technological information."(49)

Regional Action and Cooperation Between Developing Countries

Besides an international information center, UNCTAD IV also proposed the setting up of subregional, regional, and interregional information centers for the exchange of information among themselves, and between the international and national centers, the harmonization of all explicit and implicit technology policies, and monitoring of the implementation of a future international code.(50) While these centers can be of invaluable assistance, particularly to the less developed of the Third World countries, the social structure and context in which technology transfer takes place may limit their usefulness for the following reasons.

First, not only are metropolitan suppliers likely to object to the exchange of "confidential" information, and to restrict its diffusion during and after the contract period, but the local private recipients (even when not so contractually constrained) may be loathe to supply the required information for either tax reasons, fear of a regional rival, or concern about potential extra-regional competitors. Similar problems may arise with respect to the UNCTAD (supporting paper) proposal for the setting up of a system of technological cooperation on a sectoral basis in selected areas such as 1) fertilizers; 2) processing, canning, textiles and footwear; 3) manufactured metal products, electrical and electronic goods, commercial transport equipment, and petrochemicals; and 4) pharmaceuticals.(51)

Second, Third World capitalists do not necessarily have any scruples about exploiting one another. Guyana has even experienced a case of a Third World enterprise supplying supposedly 'new' buses containing reconditioned engines (which the supplier had, in turn, acquired from a developed country). Nevertheless, purchase from a Third World source (by stimulating the latter economy) is better than purchase from a developed economy, ceteris paribus. The terms and conditions imposed by Third World suppliers(52) are not much different from those imposed by suppliers in developed countries. (Underdeveloped countries should thus set the example by immediately proscribing onerous and restrictive practices among themselves.) It may be necessary, therefore, at least in the initial stages, to concentrate on the exchange of information between public enterprises in the various Third World countries. For example, there is no reason why Mexico cannot effectively transfer petro technology to Trinidad.

Third, since the UNCTAD IV advocacy of the common market framework as a means of furthering the process of technological collaboration among underdeveloped countries,(53) there has been a certain degree of regional disintegration. Both the East African and Andean common markets are very near to collapse, partly due to ideological differences, as reflected in approaches to foreign investment and transfer of technology policies. The situation is no different in the Caribbean. In addition to recent disintegration, there has always been market fragmentation caused by export restrictions. Of 15 Guyana, 32 Trinidad, and 53 Jamaica contracts in which exporting is mentioned, there is strict prohibition in 5,5, and 12 instances respectively, and exporting allowed to only certain parts of the Caribbean in 5,10, and 4 instances respectively.

In 1969, an attempt by the Caribbean Community Secretariat to introduce a foreign investment code for the region came to nought, mainly because of objections by the largest member territory, Jamaica, at the time ruled by a rightist regime which felt that the then huge capital inflows would have dried up. A second attempt, in 1974, to introduce an 'Agreement on National, Regional, and Foreign Investment and the Development of Technology' failed because the very small member territories raised objections on the grounds that they do not have either natural or human resources like the larger and more developed members and, therefore, are more dependent on inflows of foreign capital and technology. As a result, the Caribbean has not yet introduced a formal regional technology code (nor has it been possible informally to rationalize and reinforce technology strengths in the region, and minimize technology weaknesses). This regional inaction, in turn, has delayed the introduction of individual national codes, since any one member territory would not want to have measures which deviated too fundamentally from the basic elements of the regional practice, or give an undue advantage to the other members of the common market. Within the last year the integration process has almost come to a halt because of differences between petroleum-endowed Trinidad and the other members. Besides this economic polarization, there are balance-of-payments and ideological tensions between leftist-oriented Guyana and Jamaica and the others.

National Action

Because of difficulties at the international and regional levels, the time is probably now ripe for the Group of 77 (i.e. all underdeveloped countries) to agree to the introduction of a common minimum technology code by a certain date, irrespective of whether an UNCTAD-sponsored internationally accepted code (legal or voluntary) materializes in the near or distant future.   The developed countries have been employing delaying tactics while they work out a counter-strategy to the Third World attempts at building up an indigenous technological capability. In fact, whether the developed countries formally agree to the code or not, their supplier enterprises will still try to avoid and evade its measures via devious means with respect to fiscal, monetary, and other legislation. With a strictly observed Third World Code, it will then no longer be possible for supplier enterprises to play one recipient country off against another, or be necessary for the latter to attempt to outbid each other.

In this respect, and for cost effectiveness and completeness, other harmonization measures (relating to taxes and tariffs) ought to be introduced. A technology supplier is encouraged to drive a hard financial bargain and to impose certain restrictive practices when it knows that the local enterprise is in receipt of generous fiscal allowances. For example, a Venezuelan supplier of candy manufacturing technology to a Trinidad enterprise reserved the right to terminate the agreement if the local recipient did not succeed in obtaining pioneer (tax concessionary) status within six months. The recipient, in turn, is willing to make concessions because of these very fiscal incentives (e.g. tax holidays and accelerated depreciation allowances on overinvoiced inputs) and the protected nature of the market. Underdeveloped countries collectively need to abolish the old tax incentive system and buttress this with effective price control. If a new tax incentive system is at all necessary, it should be based on indexes such as volume of production, quantity of exports, and increases in employment.

The national technology center, or technology unit, would be expected to work in close harmony with a country's economic planning commission in rationalizing all the nation's policies. In the final analysis, the responsibility for building up indigenous technological capability rests with the individual state, rather than with any regional or world agencies, even though the latter can be of invaluable assistance in furthering the process, given the uncompetitive international market structure.(54)

SECTION 5: SUMMARY AND CONCLUSION

The role of technology has been made explicit in the accumulation process, and in many cases deliberately disembodied from capital for commercialization purposes. Until very recently, commentators tried to give the commercialization process analytical treatment separate and apart from the production process and the social objectives. However, technology is now being recognized as a social category which, given the perverse conditions of its importation, and the nature and context of its application, can fail to satisfy the basic needs of the community by being inappropriate and/or dysfunctional, and tending to perpetuate dependence and underdevelopment.

The market for technology is imperfect and unequal, and the profit-maximizing guest of all capitalist suppliers makes sure that there is no breach in the system. This is compounded by the tendency of most private sector recipients to be not entirely unsympathetic to the aims and objectives of the suppliers. On the other hand, the intermediate economies, given their above average level of technological capability, have been able to modify somewhat the terms and conditions of importation of technology. This homogeneity of suppliers and heterogeneity of Third World recipients has, until very recently, hardly been recognized in the literature or by the international policy makers.

The monolithic nature of the capitalist supply system suggests that the only real alternative technology, and the only genuine alternative sources, exist in the developed socialist world. However, the Third World has not exhausted the benefits that can be extracted out of the present commercialization system, by international and regional action, or the degrees of freedom at the disposal of each country, given the required political will. There has to be continued international effort to revise fundamentally the patent system, and to introduce a legally binding international code of conduct, even though the interests of the supplier enterprises are well defended by their governments at the various political forums. In the meantime, underdeveloped countries should take immediate steps to introduce their own minimum code and appropriately harmonized fiscal and other legislation; at the same time, regional centers could ensure that these were being observed. At the national level, public enterprises are likely to be much more interested in really absorbing the imported technology and really exploring alternative sources; cooperation between Third World countries is also likely to be much greater at the public enterprise level.

The less developed of the Third World countries, given their minimal bargaining power, have a special interest in seeing introduced an international code of conduct, revised patent laws, information centers, and new and appropriately harmonized fiscal systems, inter alia. In particular, they should seek the implementation of most favored nation legislation. In a market context in which there is homogeneity of suppliers but non-homogeneity of recipients, the weakest of the latter may be asked to bear the burden of the actions of the less weak. A price discriminating monopolist may charge higher prices to the least developed in order to compensate for lower prices to the intermediate economies. (An analogously perverse situation exists in which the higher prices, equivalent to lower adverse terms of trade, demanded by Arab oil producers have resulted in higher priced industrial goods from the developed world, with the bulk of the Third World countries bearing the adjustment burden in terms of both higher oil prices and higher priced industrial goods.) In terms of inter-Third World transfer action, the responsibility of the intermediate economies to the lesser and least developed economies is barely less than that of the developed world to the underdeveloped countries as a whole. Finally, changes in world commodity trade and finance are just as important as the transformation of technology relations for making progress towards the new international economic order; technology helps to create income, but technological capability requires a minimum income base for dynamic and speedy effectiveness.

NOTES

(1) For a discussion, see Commonwealth Secretariat, Towards a New International Economic Order (Final Report by a Commonwealth Experts Group) (London: Marlborough House, March, 1977), p. 54.

(2) For example, the first explicit reference to technology licensing in the Caribbean was by A. McIntyre and B. Watson, Studies in Foreign Investment in the Commonwealth Caribbean, No. 1 Trinidad and Tobago, (Jamaica: Institute of Social and Economic Research, University of the West Indies, 1970).

(3) For a discussion, see C.V. Vaitsos, "The Process of Commercialization of Technology in the Andean Pact," in H. Radice, ed., International Firms and Modern Imperialism, (Harmondsworth: Penguin, 1975).

(4) For example, the studies on Ethiopia, Chile, and Spain.

(5) For example, the studies on Sri Lanka and Hungary.

(6) UNCTAD Secretariat, Major Issues Arising from the Transfer of Technology (Geneva: Unctad, 1974). See individual monographs.

(7) UNCTAD Secretariat, An International Code of Conduct on Transfer of Technology, (New York: United Nations, 1975). For the actual code and restrictions that were recommended for abolition, see Intergovernmental Group of Experts, Preparation of a Draft Outline of an International Code of Conduct on Transfer of Technology, (Geneva: UNCTAD, May 5, 1975).

(8) For example, C.V. Vaitsos, op. cit.

(9) For example, P. Maxwell, "The Traffic in Technology," The New Internationalist, (July 1973).

(10) For an attempted quantification of the resulting dependence, see S.J. Patel, "The Technological Dependence of Developing Countries," Journal of Modern African Studies 12, no. 1 (1974).

(11) A distinction is frequently made between technology, which relates to basic knowledge, and techniques, which are used to describe the mere mechanics of applying the knowledge. For example, the production of Coca Cola in all recipient countries can easily be brought to a halt because the latter do not know the baisc recipe and only mix what is given to them.

(12) For a useful survey, see N. Girvan, ed., Dependence and Underdevelopment in the New World and the Old, Social and Economic Studies Special Number, vol. 22, no. 1, (March 1973). For an analysis of the inappropriate, product specific, and dysfunctional nature of technology in plantation economies during the precommercialization era, see G. Beckford, Persistent Poverty, (New York: Oxford University Press, 1971).

(13) Norman Clark, "Technological Dependence and Underdevelopment" (Mimeo), University of Sussex, April 1972.

(14) See, for example, Jorge Katz, Importacion de Tecnologia, Aprendizaje Local e Industrialization Dependiente (Washington: OAS Programa Regional de Desarrollo Cientifico y y Technologico. Doc. No. AC/PE-5, 1972.)

(15) C.Y. Thomas, Dependence and Transformation. The Economics of the Transition to Socialism, (New York and London: Monthly Review Press, 1974), p. 54.

(16) For an earlier basic needs type exposition, see A. Herrera, "Social Determinants of Science in Latin America: Explicit and Implicit Policy," in C. Cooper, ed., Science, Technology and Development (London: Frank Cass, 1973). For the most recent and systematic treatise, see H. Singer, Technologies for Basic Needs (Geneva: ILO, 1977).

(17) First Congress of Third World Economists, Technology Final Report, Algiers, February. 1976, p. 2 (mimeo).

(18) UNCTAD Secretariat, Technological Dependence: Its Nature, Consequences and Policy Implications. Nairobi: Fourth Session Conference, May 5, 1976 (Item 12: Main Policy Issues), pp. 20-22.

(19) In addition, UNCTAD mentioned certain objections that would be raised by technology suppliers to the effect that "private profitability is low on account of the limited purchasing power of the income groups that would consume the products" and that "the specific production of appropriate goods tailored to the unique environments of individual countries would be inconsistent with the principle of efficiency based on standardization . . . and the corporate ideology of achieving a 'global structure of excellence' based on the Western model." Ibid., p. 21.

(20) For a discussion, see F. Stewart, "Choice of Technique in Developing Countries," in C. Cooper, ed., op. cit.

(21) Junta del Acuerdo de Cartagena, Andean Pact Technology Policies, (Ottawa: International Development Research Centre, 1976), p. 21.

(22) For a discussion, see R. Hal Mason, "Strategies of Technology Acquisition: Direct Foreign Investment vs. Unpackaged Technology" (Mimeo).

(23) C. Cooper and P. Maxwell, Machinery Suppliers and the Transfer of Technology to Latin America (Washington, D.C.: Organization of American States, 1975), p. 51.

(24) M.A. Odle, "Commercialization of Technology and Dependence: The Latest Imperialist Phase: The Caribbean" (mimeo).

(25) For a discussion, see M. Merhav, Technological Dependence, Monopoly and Growth (Oxford: Pergamon Press, 1969); and D. Dickson, Alternate Technology and the Politics of Technical Change (Glasgow: Collins (Fontana), 1974).

(26) For example, according to a United Nations Secretariat estimate, in 1970 there was an average of 6 engineers and scientists per 10,000 population for 8 African countries for which data were available, 22 for Asia, 69 for Latin America and 112 for developed market economies. UNCTAD IV paper – Main Policy Issues, op. cit.

(27) UNCTAD Secretariat, Major Issues in the Transfer of Technology to Developing Countries. A Case Study of Ethiopia (Geneva: UNCTAD, June 19, 1974, p. 61.

(28) For a comprehensive listing, see Business Latin America, October 29, 1975.

(29) For a discussion, see D. Lall, Appraising Foreign Investment in Developing Countries (London: Heinemann, 1975), p. 105.

(30) For example, an early 1970s study of 56 Colombian and 15 Malaysian enterprises found that "only seven firms in Colombia and one in Malaysia paid out more than 5 percent of sales." (In the case of Colombia, the figures relate to the situation existing even before the limiting Royalties Commission legislation of 1971/2.) See Sanjaya Lall, on behalf of UNCTAD, Balance of Payments and Income Effects of Private Foreign Investment in Manufacturing: Case Studies of Colombia and Malaysia, (Geneva: UNCTAD Committee on Invisibles and Financing Related to Trade, July 3, 1973), p. 4.

(31) See Odle, op. cit.

(32) See O.S. Arthur, Commercialization of Technology in Jamaica (mimeo).

(33) For the Third World, as a whole, it is estimated that the direct costs involved in the use of patents, licenses, process know-how, trademarks, and services, will amount to $9,000 million by the end of the 1970s, a sum equivalent to 15 percent of their exports. For a discussion, see S.J. Patel, op. cit.

(34) UNCTAD, A Case Study of Ethiopia, op. cit., p. 61.

(35) UNCTAD Secretariat, Action to Strengthen the Technological Capacity of Developing Countries: Policies and Institutions, (Nairobi: UNCTAD, May 5, 1976). Item 12 — Supporting Paper, Transfer of Technology, p. 39.

(36) Ibid., p. 11.

(37) Ibid., p. 13.

(38) Ibid., p. 14.

(39) For an analysis, see U.N. Economic Commission for Latin America, Technical Progress and Socio Economic Development in Latin America: General Analysis and Recommendations for a Technology Policy (Mexico: ECLA, December 1974), mimeo.

(40) For a discussion, see C.V. Vaitsos, op. cit.

(41) Report of the Working Group, Draft Code of Conduct on Transfer of Technology (Geneva: Pugwash Conference on Science and World Affairs, April 1974).

(42) See UNCTAD, May 1976, op. cit.

(43) UNCTAD, Proceedings of the United Nations Conference on Trade and Development (New York: United Nations, 1977), vol. 1, p. 63.

(44) Even though a code is not yet in existence, concern is being expressed in supplier countries about the (little) technology that might have slipped through the net and there are recommendations for "preventive controls . . . to discourage United States technology assets from moving abroad." See J. Baranson, "A New Generation of Technology Exports," Foreign Policy, no. 25 (Winter 1976-77).

(45) For a discussion, see C.V. Vaitsos, "The Revision of the International Patent System: Legal Considerations for a Third World Position," World Development 4, no. 2 (February 1976).

(46) Especially since some of them (mostly in Africa) have become recent adherents to the 1964 WIPO Model Law which is even more injurious to their basic interests than the 1883 Paris Convention.

(47) For example in Guyana and Trinidad only 3 and 6, respectively, of the available agreements involved patents: see M.A. Odle, op. cit.

(48) See UNCTAD IV Proceedings, Vol. 1., op. cit., Appendix III, p. 178.

(49) Ibid., p. 62.

(50) Ibid.

(51) Ibid., pp. 27-31.

(52) Ibid., pp. 26-27.

(53) Ibid.

(54) In this regard, this is hardly a time for complacency (but more a time for relentless international pressure) and we find it difficult to agree with the rather uninformed government position that "not many problems relating to patent acquisition or technology purchase have so far been encountered." See Government of Guyana, The Country Paper of Guyana for the United Nations Conference on Science and Technology for Development (To be held in August, 1979) Second Draft, January 1978 (mimeo) pp. 14-15.

9 Caribbean Petrochemical Technology: Theory, Policy and A Case Study*

Steve de Castro

Almost every country in the Caribbean has claimed, at some time over the last decade, advantages for the location of some form of petroleum processing, whether or not they have either actual or potential petroleum resources. One-third of the petroleum imported into the United States passes through the Caribbean. In this sense, every island has a potential petroleum source.

This widespread apparent comparative advantage in petroleum-related activities, coupled with the absence of both a capital goods sector and any form of research into, or development of, petrochemical technology within the region, would appear to reduce the technological policy for the sector to an orthodox, but difficult, part of the prefeasibility studies undertaken for its industrial programing. The implicit assumption would be that the technology exists in, and can be transferred from, the metropolitan countries.

But there is another more abstract level at which the analysis must be conducted. To illustrate with the case of Trinidad, despite its involvement with petroleum for over a century and gas for nearly two decades, there are many senses in which all of its production, refining, and its embryonic petrochemical activities can be regarded as foreign enclaves to the Trinidad economy. Thus, to observe that the existence of advanced technological processes within an economy may mean an absence of transferred technology is to begin the series of extremely difficult abstractions required before the concept of transfer can emerge.

In this chapter, we undertake two very dissimilar tasks. In the first

*Most of the material in this paper is drawn from a study recently completed by the author for the Caribbean Technology Policy Studies project sponsored jointly by the ISER (UWI), and the IDS (Univeristy of Guyana) with the help of a grant from the International Development Research Centre of Canada.

part we critically examine the literature on technological change and transfer, especially as it relates to petrochemicals, and where the Caribbean data provides counter-examples. In the second part, we attempt, as a case study, an examination of the ammonia-base fertilizer operations in Trinidad.

There is massive nitrogenous fertilizer production in Trinidad at the Grace complex, which has been operating since 1959, based on associated gas. Grace was the first to ship liquid ammonia in refrigerated tankers. Its end products are ammonia, urea, and ammonia sulphate. Further large-scale production of ammonia, through a joint venture with the Trinidad Government and based on the new East Coast natural gas, recently came on stream. Despite this already massive capacity, the government of Trinidad & Tobago (GOTT) is seriously negotiating with Amoco to collaborate in a joint venture to build another ammonia complex, consisting of two plants, each producing 1,000 tons per day. In other words, within a few years, Trinidad will become a major center of production of ammonia and its derivatives.

In the second part of this chapter we examine those aspects of the contractual and other arrangements, under which the industry was established and is being expanded, which can be grouped under the notion of a technological policy. For most of the nearly two decades of its existence, it would appear that no such policy existed; or to put it another way, the industry was a wholly-owned subsidiary of a MNC, functioning under the Pioneer Industry legislation which allowed it to operate free of income tax for almost all of the period. As such, not only was it allowed to make any technological decisions it wished, but the now normal concerns of the technology transfer literature, such as payments for patents and royalties for processes used, were entirely absent. These payments would have been considered a deduction on profits, and profits were not only tax-free but were allowed to be freely repatriated.

The main content of our method will be an analysis of three documents:

1. The letter from Grace to the Trinidad government in 1958, which appears to be the only document which represents a formal agreement between the company and the government at the start-up of the industry;

2. the "Tringen Agreement," which is, in fact, a management contract between the Owner (the Trinidad Nitrogen Company, which is in turn jointly owned (51 percent/49 pecent) by the government) and W.R. Grace & Co., and the managing company, Federation Chemicals Ltd. (Fed Chem) which is, in turn, a wholly-owned subsidiary of W.R. Grace & Co. and registered in Bermuda;

3. a draft of the Formation Agreement between GOTT and Amoco for the Establishment of an Ammonia Plant.

We also conducted interviews with the President of Tringen and with members of the government of Trinidad & Tobago's Coordinating Task Force which is responsible for the many, heavy industrial projects being proposed for the island.

THE METROPOLITAN TRANSFER LITERATURE AND THE CARIBBEAN DATA

Presumably, one should start with some general concept, such as technological change, since the transfer notion seems to be a particular form of such change. One way these ideas could be juxtaposed is in the approach which argues that technology transfer is concerned with shifts over space of a given technique, while technological change is concerned with shifts over time. Thus, whereas one can conceive of technological change as the study of the way new techniques get generated at the frontiers between new knowledge and new production methods, technological transfer is the study of the diffusion process by which these new methods get dispersed over a wider spatial area than where they were first implemented.

But it is obviously not sufficient that a technology be operated in a certain geographical area in order to assert that the technology has been "transferred." Indeed, it would seem that it is not even necessary that the technology be in actual operation for transfer to be asserted, since there may be other nontechnological reasons why the new process is not used in a particular economy or region. A good example of this notion would be the United States decision, so far, to reject the commercial development of supersonic air transport. Certainly, the diffusion literature has repeatedly indicated that late adopters were not late because of lack of information concerning the existence of the innovation, although knowledge alone is not proof of technological capability. Specification of the conditions sufficient for transfer is, therefore, the crux of the problem, and should emerge from the notion of the geographical diffusion process, which may or may not entail transfer, or even the actual use of the new technology.

There are two levels at which the analysis of such diffusion processes can be carried out, only one of which has been treated in the specific literature. The first assumes there already exists a set of enterprises operating in a given industry, in various eonomies, when the innovation occurs, and studies the rate of imitation of the leader by the followers. Mansfield(1) selected 12 innovations occurring in the United States between 1890 and 1950; the group led by Nasbeth and Ray(2) selected ten new industrial processes in the United States and six countries in Western Europe. This work has found no common theoretical or methodological approach and, therefore, no easy, or even full, explanation for the different diffusion patterns among the processes in the countries studied. None of these processes involved a petrochemical.

A second level would involve the study of the methods by which whole new industries are set up in various parts of the world economy.

This level of analysis would require the assumption that there exists a location where the industry based on an advanced technology had its origins, and would analyze the effects of movements of factors or flows of information which helped, if not caused, such industries to be adopted or rejected in other economic environments. At its most general, this level encompasses almost the entire field of development theory. Our emphasis here, though, is on the specifics of selected industries, a more decentralized approach so to speak.

In the case of petrochemicals, this location would readily be identified as the United States, with exceptions for particular products, e.g. high pressure polyethylene (United Kingdom). Unlike Western Europe, which had an established chemical industry based on coal, the United States emerged in the twentieth century with an established knowledge and application of the use of petroleum products, first as energy products and, especially in the period between the two World Wars, also as feedstocks for the manufacture of new chemical materials. One would then proceed to study the methods by which new products and processes based on petroleum came to be adopted in Western Europe, Japan, Eastern Europe, and latterly, Latin America and the rest of the developing world.

We review now some of the theoretical notions which have been used to analyze the way the petrochemical industry has evolved in the twentieth century.

To the extent that there exists a theory relevant to diffusion of technology in petrochemicals which deviates from the orthodox approach to the theory of international trade, it exists in the notion of a Product Cycle Hypothesis as formulated by R. Vernon,(3) Robert B. Stobaugh,(4) Louis R. Wells, Jr.,(5) and others. The work of the OECD Pilot Teams Project and the Gaps in Technology study(6) also seem to indicate that data for OECD countries on trade in technology-intensive industries support some kind of product cycle theory.

The approach separates products into one of three stages in the development of the production technology and market structures, called the product life cycle, thus:

1. The New Product

   Unstandardized designs, low price elasticity of demand, constantly changing inputs, high growth rates (greater than 20 percent).

2. The Maturing Product

   A certain degree of standardization of design (product differentiation may intensify), although increase of price elasticity of demand, technical possibilities for economies of scale through mass output, increase in cost consciousness, medium growth rates (7 to 20 percent).

3. The Standardized Product

Completed standardization (with specialized forms, e.g. radios –
clock, automobile, portable), well articulated, easily accessible
international market, sensitive to price, with growth rates low in
original markets (0 to 7 percent) but higher in new foreign
markets.

Using this categorization, it suggests that new products start with
their production and markets in the industrialized, high-income countries and, as the product matures, it becomes increasingly feasible, even
desirable, to shift production to the developing countries, even if the
technique of production is highly capital-intensive.

The alternative hypothesis, which centers the explanation of
international trade on the differential factor endowments of nations,
would argue that developing countries have a comparative advantage in
the production of labor-intensive commodities, which then show up in
their export figures.

The major counter-examples which the product life cycle theorists
pose are the ability of Taiwan and Japan to develop overseas markets
for standardized manufactured products. Further, in the case of
Japanese exports, one major study found them to be more capital-
intensive than the Japanese production which is displaced by imports.(7)

The OECD study, mentioned above, found that "countries other than
the United States enter trade in the products of research-intensive
industries and in more sophisticated products in general, at a stage
where original innovation as a factor in competition is, relatively
speaking, less important. In short, trade performances can apparently
be explained by the flow of technologies, a large proportion of which
originated in the U.S."

Stobaugh(8) applied the hypothesis specifically to petrochemical
plants, and extended it to explain the licensing versus ownership of
technology in these plants. He narrowed the definition of the product
life cycle to growth rates only, using four stages – over 20 percent, 7-20
percent, 0-7 percent, and decline. Further, his methodology was no
more than to plot the percentages of plants, according to source of
technology, against the stages of the product cycle. He found that the
share of plants using purchased technology increased from 27 percent in
Stage I of the product life cycle to 73 percent in Stage III. Similar data
were presented for the "center of control" of the plants.

His main problem is that, by the definition of the theory, few new
products will be found to be manufactured in developing countries, for
the obvious reason that the theory presumes the new products are first
introduced in the industrialized countries from which they are first
imported, to be followed later by local production, and even possibly
exported back to these parent countries. For example, only 44 out of
the 350 plants he studied (12 percent) were in less developed countries.

And this is precisely one of our main problems with the theory – the
source of the genesis of new products. Because, presumably, the
concept is neutral to the stage of the product's development. That is,
one can have new products which are raw materials, intermediate
goods, or even capital goods. For example, a lot of petrochemicals are

used in products which are not new, in the sense that they have been manufactured before but from other materials, e.g. paints, detergents, dyes, and most plastics. Another problem is that for certain petrochemicals, almost all of their output goes into motor vehicles, e.g. ethylene glycol and styrene-butadiene rubber. Even if one conceives of the automobile as a new product (although surface transport is obviously as old as the hills), one certainly cannot conceive of rubber as new.

Thus it is extremely difficult to define what exactly makes a product new. Polli and Cook(9) in their tests of the validity of the hypotheses, pointed out that changes in sales of a product can vary according to the very definition of the product. To deal with this problem, they distinguished three notions – product classes, product forms, and brands. Product classes include all objects which are substitutes for the same specific needs. But in examining the car example, they conceded that need specification must entail some personal judgment. Product forms are finer groupings of product classes and include objects which, though not identical, are physically quite homogeneous. At no stage did they attempt to make explicit their conception of what constitutes a new product class or product form. In our case, petrochemicals would not be considered even a product class since they consist of a wide range of products, most of which are certainly not substitutes. Their only common bond is that they are derived from a single main material – petroleum.

Of course, it is precisely the point at which the product is new that monopoly powers are likely to be strongest. This led Nadal(10) and Cooper(11) to criticize the product-cycle theory for its inadequate treatment of the methods by which monopoly advantages in technology are preserved, or lost, during these earlier stages in the cycle. This criticism is different from our own and is a bit unfair since new products, however defined, are not necessarily monopolies or oligopolies. Two examples, at polar extremes of newness, are the automobile and the filter tipped cigarette. But a lot of new products are monopolies, at least for a time, and as such, that literature should yield much insight into the mechanisms for maintaining these powers. We do not have the resources in this brief review to look at this important topic.

Although he does not seem to be aware of the product-cycle theory, Hufbauer's testing(12) of three hypotheses for explaining international trade, in what he calls footloose synthetic materials, is perhaps the most rigorous indication so far that trade in these materials requires some other explanation than the Heckscher-Ohlin factor proportions theory.(13)

Using econometric methods, he tested the following hypotheses:

1. The factor proportions account which explains trade in terms of differential factor intensities in the best techniques of production, coupled with differential factor endowments.

2. The scale economy account which asserts that trade "arises from economies of scale, static or dynamic, which are harvested to a

different extent by each nation owing to variations in home market size."(14)

3. The technological gap account, attributed to M.V. Posner(15) explains trade as a function of the difference between demand lag and imitation lag for, respectively, the consumption and production of a new product.

Here is not the place to go into the subtleties of the differences among these three hypotheses, but it is fairly obvious that the third bears the closest resemblance to the product-cycle theory. Hufbauer's conclusions were that, for "synthetic materials," the factor proportions theory was definitely not relevant, while combinations of the other two were required to explain the variations in his data.

Even if one simplifies the problem and confines the discussion to consumer goods, one is in trouble. David Felix(16) imbeds the problem of how the new product enters the preference functions of consumers in developing countries in what he calls the International Demonstration Effect (IDE). He suggests that the orthodox utility theory is unable to deal with new goods and new variation of old goods, and he calls for an alternative theory using concepts from social psychology, such as peer group emulation, status symbols, plastic preferences, and advertising. Kelvin Lancaster(17) made some attempts to develop a consumption theory which avoids having to return to the consumer for a revised preference ordering each time a new product is introduced into the basket, but no further development of this approach seems to have occurred over the ten years since he introduced the idea.

The Hypotheses and the Caribbean Data

Bringing the discussion to the Caribbean data, it would seem that both the lack of scale economies on the home market and the technological gap notion are the inhibiting factors. Yet the Texaco cyclohexane plant in Trinidad is a counter-example to the former and was an anomaly in Stobaugh's data. It was the only one of Stobaugh's 350 sample of plants which was built in a less developed country without a large domestic market for the product. Further, the second Grace anhydrous ammonia plant in 1964 was the world's largest at the time, and the first to liquefy and ship ammonia in large refrigerated tankers.

Thus, we have counter-examples to show not only that it is possible for new technology to be mobilized initially in a developing country, but also that the export market can be used to generate economies of scale.

That it was a multinational corporation in both instances which was the executing agency is not a sufficient explanation of the Trinidad counter-examples. Certainly, Stobaugh's explanation of his counter-example (our first above), and which was also his general conclusion, was that low-cost raw materials, rather than low-cost labor, will become the important cause of international trade in petrochemicals.

He claimed that "countries with low-cost raw materials will become exporters of petrochemicals to countries with high-cost raw materials, regardless of the sizes of the respective markets of the two countries."(18) If this conclusion is correct, then institutions other than the MNCs would be capable of achieving the same export performance, at least to the extent that the entity which controls the natural resource becomes the effective overall center of control.

This explanation of Stobaugh for his counter-example constitutes a new hypothesis pulled almost from mid-air, with little intellectual justification. It was never suggested before in any of the analyses of the various alternatives, and was introduced boldly with no specification of the conditions which would cause the complete overthrow of the hypothesis he had so carefully studied.

It is not self-evident to us that the possession of such natural resources would automatically confer on its site a crucial attractiveness for the location of the many stages of downstream processing. There are still too many other variables to be accounted for in the analysis before such a simple projection of events can be made.

We are not arguing that the study of the effects of the location of a new nonproduced factor of production, natural resources, or raw materials should not be undertaken. On the contrary, we think it is vital to the analysis of any policies which Third World countries can pursue. But we are asserting that the existence of such natural resources in a particular location is unlikely to be a sufficient explanation for the development of new locations for petrochemicals and other forms of raw material processing.

Certainly, if we were to postulate the withdrawal of the services of the MNC from the two petrochemical operations in Trinidad cited in the counter-examples, and especially those with technological content, it would be difficult to envisage the initiation of these facilities.

In order to substantiate this statement, it would be necessary to examine in some detail the conditions under which the subsidiaries in Trinidad were able to obtain access to, and to apply, the parent's technology. We do this below only for the fertilizer operations.

A CASE STUDY: THE TRINIDAD AMMONIA OPERATIONS

Historical Background to Trinidad Fertilizers

The nitrogenous fertilizer industry in Trinidad is now almost two decades old. It went on stream in 1959 with a 100 tons/day (35,000 tons/year) ammonia plant, designed and built by Kellogg, using associated gas, large quantities of which were being flared or vented. The entire ammonia output was converted into 22,000 tons of urea and 70,000 tons of ammonium sulphate. These capacities were geared for the total West Indian market for these fertilizers, and hence the name of the W.R. Grace subsidiary which owned and operated the complex, Federation Chemicals Ltd. (Fed Chem).

The West Indian Federation was unable to achieve any form of common external trade policy, and Grace soon found that the plant was unable to compete in these West Indian markets with nitrogenous fertilizers from outside the region, mainly because of the relatively small size of the ammonia plant but also because of some amount of dumping. The federation was soon dissolved with apparently little further possibility for protected access to the regional market.

But Trinidad still had large quantities of associated gas being flared, and the government was willing to grant further tax holidays and other concessions for new fertilizer plants, even though these no longer would be pioneering ventures. Thus, in 1964 Grace put on stream a second ammonia plant of 615 tons/day capacity, designed and built by Chemico, to be followed in 1966 by an even larger, 750 tons/day, Braun unit. At the time, these were the world's largest single train units and Grace was the first to ship ammonia in refrigerated ocean-going tankers.

Since then, several modifications and improvements have been implemented in this complex, drawing on the expertise of the process engineers on site, and at Grace Agricultural Chemicals headquarters in Memphis, Tennessee. First, the Chemico plant's capacity was increased to 700 tons/day, mainly by operating the plant at slightly different settings than the design specifications. Subsequently, in 1970-71, it was converted to produce hydrogen, which was sold to Texaco's refinery for use in its new desulphurization operations made necessary by the increasing antipollution requirements of the United States market.

Similarly, the Braun plant has been retuned to run at up to 850 tons/day. The current ammonia capacity in the Fed Chem operations is 100 + 850 = 950 tons/day, an output which would be well in excess of the current regional demand for nitrogeneous fertilizers.

The Fed Chem Agreement

It would, in fact, be a bit of an overstatement to describe as a contract the letter from Grace to the Trinidad government (a copy of which the government signed and returned), since none of the points specified contain any references to sanctions, penalties, or arbitration procedures should either party fail to fulfill their part of the agreement, or should there be any dispute about such fulfillment.

It lists first, under 18 headings, what the government undertakes to do. This is then followed by a list, under five headings, of what the Company undertakes to do.

A brief summary of the government undertaking is as follows:

1. to grant Pioneer Status to the company to produce nitrogenous fertilizers and to refrain from granting others such status for a five-year period. The fiscal and other concessions which such status implies are well known in the Caribbean literature;

2. to provide water and electricity from the state-owned agencies at quantities and prices to be negotiated;

3. to request and recommend to the West Indian Federation and British Guiana that these territories provide protection from dumping in their markets for nitrogenous fertilizers and to provide such protection in the Trinidad market;

4. to allow the company to bring from abroad, from time to time, such skilled personnel as may not be available from the population, and to consider granting them certain fiscal concessions, e.g. duty-free importation of their household effects and relief from income tax.

The main undertakings of the company were:

1. to construct a plant to produce 100 tons/day of ammonia, and conversion to at least two types of fertilizers;

2. to acquire locally as much of the raw materials and other intermediates as possible;

3. to establish long-term training of "artisans, mechanics, and office workers from the local population, but also eventually, and to the extent possible, certain senior staff personnel";

4. to expand the enterprise "as circumstances warrant";

5. to respect the government's policy of free democratic trade unions and of promoting free collective bargaining in determining conditions of employment.

It is obvious that the technological policy of the government, implied in this agreement, was to allow total discretion of the company as to its choice and source of the technology embodied in the complex. Further, any costs incurred in obtaining such technology (such as royalties, trademarks, etc.) would automatically be allowed by the Pioneer Industry legislation to be deducted from profits and remitted abroad. Profits, in any case, were not subject to tax for the first ten years of the operation of each unit.

The Tringen Contract and Its Background

Grace approached GOTT, in January 1973, with a proposal for a joint venture to build two plants, each capable of producing 1,000 tons of ammonia per day, offering GOTT a 25 percent initial equity, to be increased at 2.5 percent per year over ten years to a 50:50 equity participation. Two GOTT chemical engineers, one of whom was Mr. Jones, were sent to Grace Agricultural Chemicals Division in Memphis, Tennessee, to look at the cost estimates of the plant and at the preliminary work Grace had done.

In March 1973, Grace proposed that the two parties go to Kellogg, the major contractor in large-scale ammonia plants, to obtain a definitive cost estimate for the two plants which were intended to come on stream one and a half years apart. Engineer Jones, along with various Fed Chem and Grace technologists, spent several short periods (less than one week at a time) at Kellogg in Houston, Texas, for briefings on several environmental variables, from local weather conditions to the chronic shortage of fresh water for cooling.

But within two or three months of the exercise, it became clear to the clients that Kellogg had allocated an extremely low priority to their project. For example, in this period the project manager was changed two or three times, each one a bit more junior than his predecessor. Further, Kellogg informed them that, because of other commitments it had assumed, the project had slipped down to ninth or tenth position in their priorities from an initial second or third. Grace suggested to its partner that they should look around for another contractor.

In December 1973, Fluor, armed with the information available from the preliminary cost estimates done by Kellogg, made a presentation of their proposals to the two partners in Fluor headquarters in Los Angeles, California; and in January 1974, the decision was taken to award Fluor the design and construction contract for one plant, to be financed by a 51 percent: 49 percent joint venture, with an eventual debt equity ratio of 75 percent: 25 percent. The final cost of the plant was about $103 million, including working capital. It should be pointed out that at the time, Fluor had never tackled such a large-scale ammonia plant, although they had wide experience with smaller ones. The decision taken was based on two points – one, that Grace was a client which had long experience in operating such plants and, therefore, could contribute tremendously to the design specifications; and the second was the relatively low price asked by Fluor, largely motivated by the latter's desire to gain experience in the design of such large plants. In fact, Kellogg has subsequently sued Fluor, claiming that Fluor has incorporated some of Kellogg's licensed knowhow in their design.

The Fluor contract was a cost plus fixed-fee arrangement which required heavy involvement at every stage by the client, who had the right to approve the purchase of every piece of equipment. In this context, the client with this power was the project manager, a Grace engineer, for although GOTT decided to locate a representative (Mr. Jones) in Fluor's offices for about nine to ten months, he had no executive powers to approve or disapprove any such decisions.

Some of the major design decisions concerned were the reformer, which was bought from Foster Wheeler, and the ammonia converter, whose catalyst, and hence design, came from Haldor Topso, the Danish firm, with the actual basket and other equipment manufactured in France and Germany. All other components were procured in the United States. A major Fluor in-house design was the CO_2 stripper, a component for which they have developed a good reputation.

The Fed Chem and Grace engineers' main contributions to the design specifications were, first, to urge maximum use of either sea water or air cooling because of the relative lack of fresh water. Second, the unreliability of the local electric utility required installation of numerous time-delay switches to deal with dips in voltage, and maximum use of steam-drive units, especially in the front end of the plant, to deal with complete blackouts. And, finally, the level of spares, which would normally be a decision of the contractor, was kept entirely as a Grace prerogative, because of the integration of the maintenance of the plant with the Fed Chem complex and because of Fed Chem's own operating expertise.

Construction began in April 1975. That is, Fluor moved to the site then, as sitework had been completed previously; and, even before the plant was commissioned, they were able to withdraw all their personnel from the plant in late July, in order to cut overheads and because the Fed Chem engineers were competent to complete the run-up operation. There was quite a bit of local sub-contracting of construction (but not design), such as the structural steelwork, and the painting, and all the pipework was cut and welded by Fed Chem artisans.

With this background, we will examine now the management contract between the Owner (the Trinidad Nitrogen Company) and the Manager (Federation Chemicals Ltd.), which is a formal legal document consisting of detailed undertakings by, and conditions on, both parties, covered under 12 sections together with an appendix specifying the cost headings under which the Owner will be charged for the managerial duties carried out by the managing company, and for use of some of its facilities. It is one of the main functions of the office of the President of Tringen to vet these charges. Another is to monitor the operations, and to report to the Board, but there is no doubt that the office represents the interests of only one of the owners, GOTT.

It is our purpose here to summarize and analyze only those aspects of this management contract which have implications for our technological policy study. Many of the limitations of the relevant information provided were circumvented by the background, given above, which was acquired mainly through our interview with Malcolm Jones.

First of all, the management agreement in Section 6 empowers Fed Chem to negotiate with the engineering contractor, to approve or disapprove any subcontractor and his contract, and to evaluate technology, patents, licenses, and knowhow available from others, which involves payments of up to $200,000 for a single fee and up to $100,000 per year for any ongoing arrangement. Further, unlike many of the other clauses, it does not explicitly require that consultation with the office of the President take place before such decisions.

In Section 4 of the contract, Fed Chem undertakes to disclose to the Owner, for use only in the manufacturing operations of the project, any technical information and data required for such operations, provided Fed Chem is also allowed to disclose the same to third parties. Further, Fed Chem warrants that it had obtained permission from its parent company, W.R. Grace, to disclose to the Owner all technical information and data pertaining to the manufacture of ammonia, obtained by Fed Chem through an agreement with its parent dated January 1, 1963, as amended. We do not have access to this agreement.

Section 10 is entirely concerned with the maintenance of secrecy by the Owner of all the information so disclosed. In particular, such secrecy must "survive any termination of this agreement."

A further restriction is that the Owner must not employ as consultants any party other than W.R. Grace & Company as long as Fed Chem remains the managing company of the complex.

The main condition on termination of the contract is that the owner will have no further right or license to use the technical information which has been disclosed during its life (which is 15 years and automatically renewed every 5 years, but with the right of either party to terminate it with 3 years notice).

The Proposed GOTT/Amoco Ammonia Agreement

There is no doubt that this joint venture would be a very different operation from the Tringen plant, since it would be an entirely new installation, as Amoco has neither ammonia nor refinery facilities in Trinidad. Also, the relationship between the two partners before the commissioning of the plant is set out much more formally, presumably because of the more active participation by GOTT in the design decisions than was the case with Tringen.

It is important to get very clear the nature of the document to which we have obtained access. It is a draft of an agreement which covers the conditions on, and obligations of, both parties, for the period between the formation of the company and the commissioning of the plant. Thus, it covers that long gestation interval in petrochemical projects between the investment decision and the plant coming on stream, a crucial period from our technology policy study point of view.

The main philosophy behind the document is that the two owners would be equal partners at all stages of decision making. For example, Section IV sets up a five member Project Executive Group (PEG) which will be the unit which will disburse funds to contractors and give legal approval to designs, contractors, etc. Two members will be appointed by GOTT and two by Amoco, but the Chairman will be an Amoco representative, with limits set on the use of his vote. For example, a quorum for any decision shall be two members, one from GOTT and one from Amoco. Similarly, Section V sets up a five member Technical Advisory Group (TAG) which will have two representatives from GOTT plus two from Amoco, while the Chairman, nominated by Amoco and approved by GOTT, shall have only a casting vote, with the minority having the right to present their views to the Board, along with the TAG decision.

The main functions of the Technical Advisory Group are to advise and make recommendations to the company on matters concerning the design, construction, and start-up of the plant.

All personnel will be recruited by the joint venture, but, in the operating stage, Amoco will be allowed a maximum of 14 expatriates, including the Managing Director and the Plant Manager, for a period of up to five years. During the construction phase, up to 38 Amoco people will be given work permits for short periods.

A Critique of the Technological Arrangement

There is no doubt that as far as the technology policy of the Trinidad government is concerned, there has been little change over the 20 years it has been dealing with the industry, and in at least one instance there has possibly been retrogression.

For whereas the first contract with Fed Chem involved no direct investment by the government, the second requires massive sums of public money which are being mobilized by the multinational corporation, which also has a virtual carte blanche on technological decisions. Similar sums are envisaged to be invested in the proposed GOTT/Amoco joint venture which, as we have seen, will be even more dependent on metropolitan technological expertise.

Further, it is difficult to conceive of the second contract as an arms-length agreement, since the managing company is a wholly owned subsidiary of one of the joint owners, namely Grace. From our information, it would appear that no attempt was made by the government to obtain alternative bids for the management contract, as it seems that this was a condition for Grace's investment, i.e. that its subsidiary should be the managing company. In fact, since so much of the costs are within the control of Fed Chem, it is difficult not to conceive the government's investment as equity in Fed Chem rather than as a joint venture.

Even in the area of technical services, where the absence of an agreement between Fed Chem's parent company and Tringen appears to be a progressive omission, there are problems because of the technical agreements between Fed Chem and Grace which are not required to be monitored by Tringen, and may very well not be monitored by any agency of the Trinidad government. It is not easy to gain access to the interface of the 20-year-old relationship between the Trinidad government and Fed Chem.

Another area of possible retrogression is the apparent omission of any explicit clause requiring the managing company to undertake the long-term training of technical and other personnel for the enterprise. It is possible that this condition may have been granted, since most of the present staff of Fed Chem are local or West Indian.

This is not sufficient to establish that Fed Chem did undertake adequate long-term training programs, as was required in the 1958 letter. Besides, other institutions may have trained people who were subsequently recruited by Fed Chem.

Whereas, on paper, the GOTT/Amoco arrangements would appear to be more progressive, it is hard to see any real differences in practice. GOTT still has no expertise in either ammonia plant design or its operation. To take the design case first, because of the structure of the Trinidad, and even the CARICOM, economy, the joint venture will be seeking out the same metropolitan firms who will procure their components in exactly the same way.

Further, because of Amoco's own lack of expertise in operating ammonia plants in the local environment, the joint venture will be relying even more on the metropolitan contractor's expertise than

Tringen did. Thus the design and construction contract is likely to be much more of a turnkey specification.

It is, therefore, quite clear that under these circumstances it is unlikely that a disaggregation of the technological package would yield many decisions different from those which might be made otherwise.

On the other hand, an investigation should be made of the local and CARICOM design and construction expertise in order to see whether certain peripheral functions could not be undertaken locally. We are thinking particularly of the foundations, the structural steel work, tanks, piping, and certain electricals. In other words, whatever Fed Chem expertise was used in the construction phase of Tringen should now be mobilized in the GOTT/Amoco joint venture, along with any other design expertise which may be available in the region.

It is, of course, difficult to envisage the institutional framework under which this Fed Chem expertise could be mobilized in the new venture. One supposes that it may be possible to work out a once-and-for-all fee, but it is obvious that Fed Chem would not be highly motivated since it would only see the new operation as a competitor. The other joint owner of both Tringen and the new venture, namely the government, should have no such conflict of interests. In fact, the more one examines these issues the more one becomes convinced that a superior GOTT strategy might have been to nationalize Fed Chem in the first place, certainly as a method of mobilizing its expertise in the expansion of the highly desirable gas-intensive fertilizer production. In any case, we have not concerned ourselves here with too many of the other issues which would have been relevant to any assessment of such a wider policy alternative, e.g. marketing arrangements and sources of loan capital, with or without nationalization.

But it is possible now to outline some ideas which perhaps could help in technological policy formulation towards future fertilizer, as well as any natural gas-intensive projects, such as methanol. For, despite Tringen's massive addition to Trinidad's fertilizer capacity, the large volumes of natural gas still unutilized will continue to attract proposals from MNCs for further fertilizer production.

First of all, it would seem a minimum requirement that management contracts be arms-length transactions between owners and managers. We do not see why such contracts cannot be the subject of international bidding, just as is the case of construction contracts. Similar technological secrets certainly are involved in which a contractor gains temporary access to a third party's technology in order to construct a plant.

Second, a detailed examination of the technological content of all construction contracts should be undertaken. Contracts proposed between contractors and third parties for knowhow should be scrutinized by the government for, inter alia:

1. biases in the source of the technology;

2. possible unnecessary payments, e.g. patents may have expired;

3. potential overcharging on royalties, etc.;

4. potential capital goods import substitution, and compatibility with existing plants for spares, maintenance know-how, etc.

Third, all fertilizer plants should be required to maintain a central pool of spares similar to the way the sugar factories operated, in order to maintain minimum plant downtime and to reduce the scarce capital and foreign exchange tied up in spare parts.

Fourth, all fertilizer plants should be required to contribute to the maintenance of a single central laboratory for the monitoring of existing processes both here and elsewhere, and for any trouble shooting activities such as a prematurely poisoned catalyst. A detailed examination should be made of the potential of this central laboratory for undertaking a selected program of R&D type activity in the field of nitrogenous fertilizers and any related petrochemical processes, such as methanol.

And finally, a program of training of young engineers and technologists should be lauched, in which they are rotated among the different plants and exposed to as wide a range of experience in as short a time as possible.

NOTES

(1) E. Mansfield, "Technical Change and the Rate of Imitation," Econometrica (October 1961); and E. Mansfield, Technological Change (New York: Norton, 1971).

(2) L. Nasbeth & G.F. Ray, eds., The Diffusion of New Industrial Processes. An International Study (Cambridge: University Press, 1974).

(3) R. Vernon, ed., The Technology Factor in International Trade (New York: National Bureau of Economic Research, 1970).

(4) Robert B. Stobaugh, "The Neo-Technology Account of International Trade: The Case of Petrochemicals," in Louis T. Wells, Jr., ed., The Product Life Cycle and International Trade (Graduate School of Business Administration, Harvard University, Boston 1972).

(5) Louis T. Wells, Jr., ed., The Product Life Cycle and International Trade (Division of Research, Graduate School of Business Administration, Harvard University, Boston, 1972).

(6) OECD, Gaps in Technology Between Member Countries. Analytical Report, Paris, 1970.

(7) M.Y. Yoshino, Japan's Multinational Enterprises (Cambridge: Harvard University Press, 1976, p. 17).

(8) Robert B. Stobaugh, The International Transfer of Technology in the Establishment of the Petrochemical Industry in Developing Countries, UNITAR Research Reports, no. 12, United Nations Institute for Training and Research, 1971.

(9) R. Polli and V. Cook, "Validity of the Product Life Cycle," Journal of Business 42, no. 4 (October 1969).

(10) Alejandro Nadal, "Multinational Corporations in the Operations and Ideology of International Transfer of Technology," Studies in Comparative International Development 10, no. 1 (Spring 1975): 15.

(11) Charles Cooper, "Science, Technology and Production in the Underdeveloped Countries: An Introduction," Journal of Development Studies 9, no. 1 (October 1972).

(12) G.C. Hufbauer, Synthetic Materials and the Theory of International Trade (London: Duckworth, 1966).

(13) See Bertil Ohlin, Interregional and International Trade (Cambridge: Harvard University Press, 1933).

(14) Hufbauer, op. cit.

(15) M.V. Posner, "International Trade and Technical Change," Oxford Economic Papers 13, no. 3 (October 1961).

(16) David Felix, "Technological Dualism in Late Industrialisers: On Theory, History and Policy," Journal of Economic History 34 (March 1974): 194-238.

(17) K. Lancaster, "Change and Innovation in the Technology of Consumption," American Economic Review 56, no. 2 (1966): 14-23.

(18) Robert B. Stobaugh, op. cit.

IV

Science and Technology Policies in Latin America

10 Organization of Scientific and Technological Development in Latin America

Jaime Lavados

THE INSTITUTIONAL EVOLUTION

During the decade of the 1950s, Latin American recognition of the role of science and technology in socioeconomic development, though vague, was beginning to become widespread. Soon organizations destined to promote and improve scientific and technological activities began cropping up. This happened, initially, in those countries which previously had more of a scientific tradition (Argentina, Brazil, Chile, and others), and later spread.

Institutions with these objectives had long existed, particularly in those countries indicated above, reflecting the presence of a certain number of scientists and research institutions. In this respect, certain organizations should be mentioned, such as the Academies of Science, Engineering, and Medicine, Organizations for the Advancement of Science, and scientific societies. Also of great importance are those universities which, from the onset, fully recognized the importance of the function of scientific investigation as one of its most important tasks.

On the other hand, those organizations that turned up during the 25 years of the postwar era have certain characteristics which differentiate them sharply from those established earlier. Due to recognition given the usefulness of science and technology in economic development, governments have been the agents that, together with preexisting scientific communities, set up appropriate organizations and provided the resources for them to conduct research. The structure of these organizations reflected the model prevalent in European countries, and UNESCO played a part in its diffusion to Latin America. They are composed of active scientists and a few technologists at the highest level available in the country, who are appointed either as individuals of merit or as representatives of the President of the Republic (or both). The institutions appear to be directly tied to the President for their substantive activities, even when, administratively speaking, they are tied to the state through the Ministry of Education. They are called the CONACYTS, an acronym for Consejo Nacional de Ciency y Technologia (National Council of Science and Technology).

Planning was a method used widely throughout Latin America about the same time. It was conceived as a useful means of defining policies and plans for the development of a scientific and technological capability. More traditional activities of the old institutions (such as conventions, seminars, and publications) have been deemphasized even though maintained in practice.

In the statutes and statement of principles, the CONACYTS are defined as being responsible for the nation's global scientific and technological development. However, due to their limited resources, their lack of jurisdictional and administrative articulation with other state operative agencies and, above all, their own perception of function, in practice the CONACYTS are restricted to developing activities aimed at obtaining the following specific objectives:

Strengthening of National Scientific Capabilities

Generally, this is done through financial aid to research projects; the granting or negotiating of scholarships and other facilities for continued training of scientific personnel inside the country or abroad; improving scientific facilities through purchase of equipment, or through installation of well-equipped laboratories; development of libraries; aid in starting and/or participating in scientific meetings, and other such activities.

Previous Studies for the Design of Plans and Programs for Scientific Development

Generally, these "base studies" have consisted of inventories of the scientific and technological capacity available (personnel, financial aid, research units, institutions, and scientific projects in progress). Also, they have conducted studies of specific partial aspects of importance to the problem, such as the transfer of technology (in its different aspects), diffusion of scientific knowledge, the role of diverse agents in the diffusion and use of scientific knowledge, and budget programing in science and technology.

Design of Policies and Plans for Scientific and Technological Development

Plans have been oriented to the detection of those scientific areas – rarely to technological areas – which demand more development, either because of present comparative weakness or because of potential social and economic importance. Objectives are defined in reference to formation of personnel, costs, and some other variables. Occasionally, more general programs are proposed, that is, not limited to one area or sector, but aimed at solving problems which reach all, such as information systems or education for science.

These plans and programs have not generally included the operational details which assure the achievement of the established objectives. They have consisted rather of global proposals sent to the President of the Republic, or to other officials, and which have rarely, if ever, been even partially implemented. On the other hand, the CONACYTS have not had the executive power to attempt achievement directly. The planning task has, therefore, become a lateral and distant one, and the CONACYTS have concentrated instead on the management of their own scarce resources.

Yet it is not merely this defective insertion into the structure of the state, nor the quality of its plans, which limit CONACYTS' planning role and explain its limited economic and political support. There are also other factors, among which two should be pointed out: its composition and origin, and the conceptual and operative expansion which has taken place in the last few years, both in reference to what the scientific and technological system is understood to be, and what the actual boundaries of science policies are.

The central thesis of this work rests largely on this last point and it will be more fully developed.

In reference to the origin and composition of the CONACYTS, it is worth noting that they were born first in the countries where there already were previous scientific communities. In addition, in practically all cases, the initial impulse to establish CONACYTS came from proposals of scientific groups "acceptable" to the governments. Rarely, if ever, were they the result of political decisions by the central power. The role played by UNESCO in the establishing of these institutions has also been important. These situations explain their composition — scientists of the highest level — and the organization model adopted, both of which generated important consequences for their orientation and basic work objectives.

It is natural for active scientists to conceive that the most important role for an institution is the facilitating and stimulation of scientific research of high quality. They may concede that it is preferable that such research be carried out in areas, and in reference to problems, of special importance to the country. Certain steps may be taken to orient the funds transferred and the personnel involved in that direction. Yet, their main concern is directed more toward the technical quality of the work than toward the area or problem being investigated. They are not concerned with the dissemination of knowledge beyond the scientific community or with the needs of higher education in general. They do not consider the problems linked with the productive utilization of the scientific and technological knowledge to be their responsibility.

Because of this, it is understandable that about the same time, and without any relationship to the work of the CONACYTS, other institutions dedicated to the creation, diffusion, and use of knowledge, or to preparation of technical personnel, appeared in Latin America.

These are agencies clearly oriented toward concrete areas or sectors.

Institutes of applied and/or technological research appear directly linked with certain problems of varied socioeconomic sectors (agriculture, manufacturing, mining, fishing, and forestry). In Chile, alone, the number of such institutes rose to 18 in 1970. Some of these have even more resources with which to work than the CONACYTS. They are created by independent decisions – not coordinated in any high level plan – by the Ministries or other agencies for the economic and social development, as was the case of CORFO in Chile).

Certain countries have developed centers of measurement and normalization that not only provide an important challenge but, under certain conditions, a technological stimulus to the productive sectors. Consulting firms in engineering and technological information services have been developed. Finally, certain state agencies of technical assistance have been established.

Universities have become involved, not always successfully, in the management and organization of scientific and technological activities, and, ultimately, have started exploring more appropriate mechanisms for research hiring of, or contracting for, certain technical services, with the productive sectors.

On the international level, the scientific variable and, above all, the technological one are incorporated into the discussions, and even become part of the concerns or policies agreed upon by groups of nations. Current discussions by the Central American Common Market, and the decisions on industrial property and the transfer of technology of the Andean Treaty, are eloquent examples.

Also, increasingly, scientific and technological development and its application to national production needs have become the subject of research. Several studies, among them "Base Studies" of the CONACYTS or works of international agencies, have attempted to redefine and to broaden the scope of the Scientific and Technological Policy. The specific work of other institutions has stimulated this redefinition. Several problems of concern from an economic, social, and political point of view have been identified, the number of agents recognized as having influence in the scientific and technological development increased, and the factors that limit the productive application of goods and services are minutely analyzed.

THE CONCEPTUAL CHANGES AND THEIR CONSEQUENCES

This process, summarily described, has induced the appearance of significant changes, not only operational, but also – and fundamentally – conceptual. The very concept of a Scientific and Technological Policy has changed. New ideas on the objectives and ends of that policy have appeared. The scope of action is broadened when the number of actors and processes involved in creating and disseminating scientific and technological knowledge is increased. The importance of education, beyond mere preparation of researchers, is acknowledged. Finally, the aim is to redefine the methods and instruments for planning and implementation of policies in this field.

The smaller Latin American countries, with a small or weak scientific tradition, do not install CONACYTS. They experience little pressure from their international scientific communities and become integrated into this process only when sufficient conceptual and operational changes have been made. Generally, they prefer to develop units of science and technology within their Planning Boards.

A crisis in scientific and technological policy and planning is visible in Latin America. To the interest in generating a scientific and technological "offering," "capability," or "infrastructure," as an almost mechanical stimulus to social and economic development, is now added an interest in knowing how to find and manage those factors which stimulate demand for adequate knowledge to facilitate acquisition of productive necessities.

The importance of utilizing locally produced or adopted technological knowledge is emphasized. This stimulates growth of local technological capabilities, makes a certain national independence possible, and promises favorable financial consequences due to the high price of foreign technology. On the other hand, this also stimulates thinking about the design and development of appropriate technology, as applied to the specific circumstances of the country (economic, social, cultural, and political).

In addition, an important turnabout in the nature of professional dedication to scientific and technological politics and planning has taken place. No longer do the most active and prestigious scientists monopolize that field of study or serve as heads of the national or international institutions which work in those fields. Now engineers, technologists, and, especially, economists are included. This change in the professional makeup is at the same time both a consequence, and a causal factor, of the changes that have occurred.

Nevertheless, expansion in the scope of science and technology policy and recognition of the large mutiplicity of processes and actors that have, or may have, influence in the scientific and technological development and in the application of knowledge in the productive activities, have other consequences. Conceptual or operational consensus on these problems, relative to more appropriate organizational solutions, have not been produced. In those same countries where CONACYTS appear and remain, other government agencies are established with the same apparent goals. Nevertheless, it appears that these new agencies are now interested in the problems of applied knowledge, rather than the improvement of local scientific and technological capabilities.

If we go deeper into the study of the processes which influence demand for knowledge, or its diffusion, and into the educational institutions which train scientific and technological personnel, we find additional complexities. The majority of these actors and agencies (productive corporations, consultants, technical services, or state agencies) have their own objectives which are not scientific and technological per se. Activities in this field are strictly limited to the way in which knowledge is perceived as being relevant to the attainment of their own purposes. It is perfectly understandable that they should tend to see science and technology as an element whose practical value is only as an instrument. Even the governments' assistance to these activities, solely because of their cultural or educational importance, is minor or infrequent. At the same time, the research institutions, the scientific communities, and certain educational centers have only a partial, or a totally different, view.

This enlarged scientific and technological community encompasses a great number of actors and processes oriented towards their own disssimilar, and even conflicting, goals. They are only bound together by the fact that they participate in, or influence, the generation, transmission, and utilization of scientific and technological knowledge. Their real behavior is determined by the pursuit of their own particular goals. Their perception of scientific and technological development will depend upon their standing within this complex system, and their respective roles in relation to fulfillment of personal goals.

In the same way, they depend upon those objectives to which a greater importance is assigned, the functions and behavior which will be emphasized, and the model of scientific and technological development will be borne in mind. Undoubtedly, many discussions of planners, administrators of science, scientists, technologists, businessmen, and international organizations in Latin America have a tendency to be partisan because the scientific and technological policies that each one supports are those which best serve his/her high-priority function.

This functional partisanship of the scientific and technological community tends to negate existing interrelationships between the different components of the system. A classic example is the international transfer of technology. When this problem is managed solely at the level of jurisdictional restrictions for the importation of technology, without at the same time taking action in the development of certain local technological capabilities, the whole productive apparatus is brought to a standstill. The underdeveloped countries cannot import what they require, yet there is no local capacity for filling local needs.

ORGANIZATIONAL WAYS AND PREREQUISITES

The need is clear for greater integration of the system. In Latin America, only the governments can do it. They should recognize the complexity of the system, and at the same time maintain an overall perspective of its functioning.

An integral view does not mean an institutionally centralized solution. It is evident that the fulfillment of diverse objectives is a task that cannot be individually accomplished by any one institution. The concept of a "Scientific and Technological System" has a great degree of abstraction. It is not something with ontological density in itself. It is only an analytical instrument which helps to visualize better the elements and processes which interrelate and have influence in the creation, dissemination, and use of knowledge. The incorporation of an actor into this system does not necessarily make him lose his identity.

In addition, decentralization is desirable to foster the diverse means necessary to develop each of the specific tasks.

The motives pointed out make it quite evident that the organization and administration of scientific and technological development in its entirety should have several concurrent characteristics:

1. Should be capable of reflecting the differences of elements and actors within the scientific and technological system;

2. Should make possible integration of these elements into a functional whole, with some cohesion and common orientation;

3. Should be able to generate or stimulate those missing or weak components of the system in order to facilitate the work of the system as a whole;

4. Should make compatible the objectives of individual actors with the wider objectives of the general society;

5. Finally, should recognize and serve the characteristics and possibilities of the specific country in which it works, in its design and operation.

Naturally, it is not possible to set forth a finite design for the organizational model for development of science and technology which would be applicable to any nation at any moment. Nevertheless, an outline could be made which contains certain principles for organization of the scientific and technological establishment that can resolve problems discussed above.

The dissimilar objectives and behavior pertaining to the individual elements that make up the system make useful a decentralized organizational solution which can correspond appropriately to those structural and functional differences. This is the line of thought emphasized throughout this chapter. Nevertheless, decentralization could mean atomization, lack of coordination, and serious difficulties in establishing objectives common to the whole system, a type of laissez faire operation historically inefficient in these matters in Latin America.

Decentralizing could also mean a multiplication of costly and bureaucratic planning and administrative institutions. This could be particularly serious in small and poor countries where sufficient economic capability, adequate personnel, and a reasonable volume of scientific and technological activities that could justify that proliferation is nonexistent. An American style solution, with its enormous multiplicity of centers and institutions which are totally independent of one another, is clearly not feasible in Latin America.

In part, this is due to the fact that our economies could not stand the cost, but also because this type of solution goes hand in hand with available human resources, business behavior, and previous industrial and technological development, which cannot, for the present time, be compared with those of Latin America.

Actually, in Latin American countries we do not now find a situation which makes possible the spontaneous and fluid functioning of a market mechanism in science and technology. Long-term programs are necessary in order to establish and develop many of the elements of the scientific and technological system. Multiple decisions and political, economic, and financial steps would have to be taken in order to generate enough demand for appropriate technology from the productive sector.

If we take these problems into account, a scheme for the organization and administration of scientific and technological development should at least have the following components:

1. Mechanisms to stimulate internal, general, scientific and technological capability;

2. Sectoral and technological management mechanisms; and

3. Internal coordination of the scientific and technological systems as a whole, and with overall comprehensive objectives and politics.

Mechanisms to stimulate scientific and technological capability is an obvious objective. It is truthfully the main goal of the CONACYTS. As we have seen, they were not qualified to develop other tasks. It is evident, for reasons previously discussed (intrinsic characteristics of the scientific work, stage of development, objectives of the actors, etc.), that the demand which emanates from the socioeconomic sectors is not sufficient to stimulate the development of the scientific and technological capability with the requisite force.

It seems desirable to reconstruct the function of this organization in order to gain two interrelated objectives: 1) to launch the development of a minimum scientific and technological capacity, generalized and balanced, from the diverse fields of the more important sciences and technologies; and 2) to orient the development of increased capacity and specialization towards specific problem areas relevant to the country.

The activities developed with these goals in mind are to stimulate the education of scientists, promote the development of high quality researchers, create communication systems taking advantage of the world knowledge flow in the sciences, be it personal (conventions, conferences, courses) or written (libraries, journals), and collaborate in the establishment of research institutes.

It is fundamental to reiterate that functioning will depend on clarity of objectives, on quality of operation, and, in a very substantial way, on the magnitude of the financial resources at its disposal and on the quality of its leaders and administrators.

Sectoral and technological management mechanism. It is difficult to promote the development of the scientific and technological capability, and then define and operate specific technological policies. Also there are differences between the actors, instruments, and specific objectives of one sector in relation to other sectors. The technological policies of a given sector should be linked with its socioeconomic policies. Therefore, sectoral mechanisms for policy and technological administration should be structured into the sector being served. This seems to be the only way of coping with the complexities mentioned. Naturally, the definition of what constitutes a sector may vary according to country, depending on its relative degree of development, its objectives and national priorities, and specific operational possibilities.

The function of this mechanism is to determine the technological requirements which the socioeconomic policies assume, and to develop an administrative and financial capacity to operate through available instruments, either in the sector itself or outside of it, yet above the actors and institutions that relate to the sector (corporation, technical, financial, or educational instruments, units of research and development, or government agencies).

Coordination of the scientific and technological system, both internally and with the rest of society. The mechanisms herein mentioned are not capable of solving some problems, thus increasing the difficulty of sectoralizing.

Among them the following should be mentioned – policies and decisions of government, which, taken into account with other aims, have an influence over the creation, diffusion, and utilization of scientific and technological knowledge (for example: tax policies, fiscal tariffs, industrial and financial measures). It is clear that the technological components of these measures would be more highly valued if some mechanisms capable of evaluating effects in the scientific and technological field exist.

Several activities which take place in the scientific and technological systems are not aimed at a specific sector. Among these are such activities as: technical information services and centers of information that, by their degree of development and by that of the national economy, become costly and inefficient to restrict solely to one area or sector. Activities of normalization and measurement are in the same situation.

A general problem that should be considered is the transfer of technology. Even though it has been linked much more with industrial development, it is also evident in mining, agriculture, infrastructure and services.

Even though sectorally decisions could be taken, and policies implemented, with respect to certain components of the scientific and technological system which are absent or weak, the same decision of setting up sectoral mechanisms (at which level, which limits the defined sector will have, etc.) could be required of a suprasectoral coordinating body.

An additional reason, of a much more general nature, suggests the design of some suprasectoral body. The general orientation of the scientific and technological system implies decisions with respect to priorities and, consequently, to the assignment of resources. A comprehensive view is absolutely necessary in these matters.

It is not possible to elaborate on the temporary profile and strategy to achieve organization. It will depend on each country, on its present institutional organization (in some cases, it will only be necessary to regroup those already in existence, in other cases, it will be necessary to create what is missing), but above all, it will depend on the political decision each country takes.

Development of science and technology presents us with such a variety of problems and complexities, especially in the present and expanded concepts of the scientific and technological systems, that little can seriously be done without a clear and maintained political decision in that direction. On the contrary, if a sustained political commitment does not exist, we will continue to be the witnesses of weak, irrelevant, and inefficient efforts which at present are felt in the majority of Latin American nations.

11 National Science and Technology Councils in Latin America: Achievement and Failures of the First Ten Years

Eduardo Amadeo

Usually expected from a paper of this sort is a more or less critical analysis of the results obtained by the CONACYTS, and some recommendations about the best ways of improving their current performance. I think, however, that the problem is more complex. It is not possible to study the CONACYTS in themselves, isolated from the complex reality where they were born and that today defines their possibilities for expansion. But I also think that it is no longer possible to analyze them with the same theoretical perspective that backed their start and that we have been using as a paradigm through the last ten years.

With this approach in mind, I first will give a look at the theoretical approach developed in Latin America during the last ten years regarding the problems of science and technology planning. In the second part, I will describe the conditions in which the CONACYTS were born and grew, and finally I will analyze the experience of four countries. In the summary, I try to present a few ideas for an alternative approach to this problem that may have more analytical and, in consequence, political relevance to our concrete problems.

THE ACADEMIC BACKGROUND

In the mid-1960s, a number of researchers from various fields of social science began to study some issues connected with the generation, transmission, and use of scientific knowledge in less developed countries, in particular in Latin America. Many years of work by economists, sociologists, engineers, and others resulted in a body of knowledge, once defined as the "science of science," which has had a deep influence on a number of political actions taken by the governments in this field. As a concrete case of successful application of these ideas, we must recall the studies on technology transfer developed by Vaitsos and the Andean Pact Group,(1) whose results

149

served as an academic basis for the adoption of severe protective measures taken by our countries against some prejudicial forms of technology imports.

Also, in the case of planning and execution of S&T policies, the actions implemented received the theoretical support of many authors and, in particular, from international organizations.

Among the contributions during these years, we have to mention an approach that, given the way it influenced the thought on these matters, can be described as a paradigm: the systemic approach.(2) Francisco Sagasti, considered the most relevant author of this approach, defines a system as "a collection of interrelated entities, each of which affects, at least potentially, the behaviour of the other."(3) According to this definition, "a nation could certainly be considered as a system."(4) The S&T system is thus defined as "a collection of interrelated entities and operations, which generates and transforms the intangible good knowledge."(5) Sagasti and other authors who have closely followed his approach offer very detailed taxonomies of the activities of the S&T system, and the interrelations among its components. This exercise is completed by proposing ways and means to improve the systematization of the whole, up to a point where it is possible to outline the characteristics of an "idealized planning methodology." According to Sagasti, a model can be constructed "regardless of the area or region for which the planning emthod is to be applied . . . (due to) some development in operations research and in social sciences that provide the tools for approaching the characteristics of an ideal planning methodology."(6)

All the policy actions suggested and supported by the Regional Program of the O.A.S. in particular, and many other international institutions, rested on the basic assumptions of this approach and utilized them as their main guidelines. Furthermore, we find a direct influence of this approach on the objectives attached to most CONACYTs in Latin America and on the general policy statements of most governments. The strengthening of the S&T system was considered as "the" key to a successful policy on this area.

During the last three years, the systemic approach has received a number of criticisms, particularly from authors who feel there is a need for conceptual schemes more suited to the complex specificity of each historical situation. Oscar Oszlak, for example, states that: "This ideal model of relations can be seen reflected in the planning proposals which conceptualized ideal relations from abstract policy definitions, not incorporating the idea of conflict."(7) "Even though such global schemes have fulfilled a necessary step for the recognition of S&T as a problem area, they have serious limitations due to the formalistic and rhetorical nature of its statements."(8)

In a recent paper written with Liliana Acero,(9) we argue that: ". . . there are three aspects that have to be taken into account if the system analysis is to be used as a paradigm:

1. It focuses on the interrelations and its results: the flows; not on the determinants of such flows. At first sight, we visualize the

subsystems as black boxes which generate products, services, resources, and receive similar inputs from the corresponding subsystem . . . Regarding the S&T subsystem an ideal view of its structure is presented as a basis for its analysis. There seem to be no social actors, definite interests, whose differential behavior or relations may explain the dynamics of the subsystems, their extrarelations, and the quantity and nature of their flows.

2. In connection with the above, the regulatory political and cultural subsystems appear as disconnected from the society they regulate, as aloof entities which emit orders, create patterns, etc., not being influenced by the characteristics and dynamics of such societies . . .

3. In this whole approach, there is a marked tendency to consider technological aspects as having some internal rationale that determines their evolution, both in the relation between its components: science, technology, diffusion; as well as with the 'outside world.' "

Another line of work that has flourished in the last years takes as a basis for its proposals the potential influence of the State's buying power in the generation of demand for S&T.(10) The conceptual framework used by this line of thought could be summarized as follows: 1) S&T are important for development, 2) the innovation process in less developed countries (LDCs) requires the allocation of vast resources which, very often, the private sector lacks, 3) the State enterprises play a key role in many important sectors of our economies, 4) the experience of other countries (e.g. France, Japan) shows the feasibility of using the State's purchasing power as an instrument of S&T policy, 5) the main restriction is of a psychological nature: the unawareness of the State officers about the importance of technology for development, 6) were such restrictions overcome, the need would be to propose adequate mechanisms for the implementation of technological policies in the State enterprises. The use of the State buying power could then be used as a complement to the action of the CONACYTS, increasing the demand for locally generated technology.

Similar to the case of the systems analysis, there is a definite tendency here towards isolating the problem from its determinants. The State is considered only as an administrative apparatus, with little reference to the interests that define the political dynamics of any society. Hence, the fact that such a passive or negative attitude of the State has any relation to the interests of the ruling sectors of the society is not taken into account. As Oszlak correctly points out: ". . . these analyses assume that in each area of the social activity there prevails a general interest that is expressed in the State, to which the social actors and institutions are subordinated. Such views assume that the state is capable of defining and setting normative frameworks to disaggregate such general interest in sectoral ends and objectives."(11)

It is worth noting, too, that the extrapolation of successful experiences from other countries, such as France and Japan, has been made without taking into account, even at a purely analytical level, the specific historical and political conditions of their development process. In this respect, we may ask whether the peculiarities of the French bureaucracy can be found in the State apparatus of our countries; or whether the conditions of political power of the Japanese industrial bourgeoisie are often reproduced in the dependent world.

With respect to the demand coming from the private sector, both approaches assume that the State should induce it, because the fact that very few local entrepreneurs are aware of the possibilities of S&T as a fundamental input for the improvement of their productive activities is an example of the underdevelopment or backwardness of the managerial class in our countries. Based on this assumption, a number of studies were undertaken which analyze the attitudes of the entrepreneur, mainly from a sociopsychological point of view.(12)

These uncritical and out-of-context perspectives have prevailed in Latin American thinking about the problems of S&T planning throughout the 1960s and the 1970s. Maybe their greatest theoretical restriction lies in the fact that they assume that this problem of the "science of science" has its own rationale and, therefore, can ignore the contributions of other areas of social sciences that could help to make it more relevant to our complex reality.

The political recipes derived from these approaches had, in consequence, their same characteristics, in particular their universalism and blind trust in the possibilities for self-expansion of scientific knowledge. In other words, whatever the political, economic, or social conditions of a country, the solution for its S&T problem lies in the establishment or strengthening of research capacity, coordinated through an institution located at the highest bureaucratic level. The knowledge generated by this structure would stimulate the latent demand and the loop would thus be closed. The possible deviations from this ideal or model, usually treated as pathologies, would be overcome through an adequate planning structure and the subsequent increase in financial support. The State could fulfill its dynamic role once the officers related in one way or another to the S&T problem were convinced about the importance of scientific knowledge . Also implicit in this scheme is the idea that the S&T element is fundamental for the consolidation of the productive sectors, and, therefore, the expansion of the S&T activities would easily find political allies to support the expansion of the system.

In the following sections, I try to qualify these models through an analysis of the concrete experiences of the countries of the region. In the last few years a number of authors began a line of analysis by questioning the most basic premises of the problem. Among these authors, a special mention has to be made of the FINEP (Financier of Studies and Projects) group in Brazil, whose work has opened new ways for the understanding of the problem, and has provided valuable ideas for an alternative institutional framework for the execution of S&T policies. Let me briefly mention those studies that more clearly show this new and refreshing view of the problem.

A group of three studies, published by FINEP within a single book(13), analyze three cases of innovation in different sectors of Brazilian industry: textile, paper, and cement, which represent three different types of oligopolistic markets. The main assumption behind the analytical scheme used in that technical progress is "neither an exogenous variable, whose effects are spread through the economy, nor an unavoidable consequence of capital accumulation, but an element that has different meanings within each market structure."(14) The main conclusion of these studies shows that:

1. In the three cases studied, the availability of "appropriate technologies" did not, by itself, assure its adoption by the firms; technologies that were clearly cost-reducing were adopted only after some structural conditions had changed;

2. Technology is not regarded as a key element for the market position of the firms in the sectors analyzed; and the "technical features" of the technology are not a determinant of their adoption. Product diversification, the vertical integration of the productive process, are presented as the main determinants of the market structure of the sectors.

3. The technologically-minded entrepreneur is not necessarily the most successful one in the sectors covered by the study.

These findings suggest to us the need for a careful review of the promotion policies and their instruments advocated by the current literature.

Working on a subject much closer to our specific interest, that of planning of S&T, Vera Maria Candido Pereira(15) tries to "emphasize the importance of the analysis of S&T planning as a political act in itself."(16) Her approach is an effort to articulate the needs of the economic level with the actions and decisions of the political level, while taking into consideration the demands, alternatives, and infra-structural limits which historical accumulation patterns place upon the choices made at the superstructural level of the State. This perspective qualifies the concept, so widely diffused among authors, that science planning can be considered as a purely administrative act which, in its technical neutrality, can be isolated from the social values and interests. With this approach, Vera Pereira develops a sharp analysis of the evolution of S&T planning and policies at different historical moments in Brazil.

These contributions, and many others that I cannot analyze here in detail, gain for the "science of science" some of the concepts developed in the Latin American "structuralist" school.

With detailed and contextual analyses of the role played by S&T in the insertion of our countries in the international division of labor, and in the interplay of social and political interests, we could answer some questions that the literature has barely touched:

- Why is it that only in the 1970s the first explicit S&T plans are elaborated in our countries?

- Why has it always been the State which was the promoter and supporter of research institutions?

- To which conjunctural articulation of social forces at the political level is associated the decision to plan these activities? Furthermore, how have these situations evolved, and how was this reflected in the evolution of S&T institutions and in the explicit social interest in science?

THE CONACYTS

A General Approach

A number of countries, especially those with a long university and scientific tradition like Argentina, Brazil, Colombia, and Peru, had established their academies on various scientific fields well before the 1960s. Brazil and Argentina had created National Research Councils during the 1950s, but the decade of the 1960s is the landmark in the formal evolution of the S&T institutions in Latin America. Uruguay (1961), Chile (1967), Venezuela (1967), Argentina (1969), Peru (1968), Colombia (1968), Mexico (1970) established new institutions aimed at the planning, coordination, and promotion of scientific and technological activities, most of them located at the highest bureaucratic level (State Secretariat or advisory councils to the Presidency).

The rise of these institutions during the 1960s is not a casual fact. Two factors explain the sudden interest in the planning of scientific activities in Latin America. The first one is the fantastic growth in the S&T budgets in most advanced countries, in particular in areas related to space and military activities and their subsequent impact on the productive sectors. The way in which this process was conducted, particularly in the United States, led to a situation of "excess of supply" of knowledge, where only those firms having the capacity to absorb the available knowledge were able to obtain oligopolistic positions in their markets. In this way, technology became the competitive factor and defined a new pattern of intercapitalistic competition which necessarily had to affect our countries.(17)

The second fact is the "planning fashion" that appeared in Latin America after the Punta del Este Conference. With the "Alianza para el Progreso," the United States offered a new pattern for the allocation of aid, which required from the recipient countries a clear definition of their financial requirements, in other words, development plans.

The confluence of the generalized faith in the possibilities of S&T, and the "need of planning," led our countries to the belief that S&T had to be planned, and that adequate institutions were needed for that purpose. These institutions adopted various formal structures, and they were introduced in different ways into the State apparatus; but most of them had the same tasks: in general, planning, promotion, and coordination of research activities; in particular, allocation of funds through grants, fellowships, support of information services, international cooperation, and the production of S&T development plans.

But there are two main characteristics of this process of planning of S&T which are worth pointing out. The first one is that the whole process is made from the scientific supply side of the society, the rationale being that an adequate allocation of funds for research would produce results, these results would be absorbed by the society, and this, in turn, would induce further demands to the S&T system. For this process to be put into motion, all that was needed was the firm belief of the officers and politicians in the possibilities provided by S&T, and an adequate institutional structure, such as the one suggested by the international organizations.

The second characteristic is the fact that, in most cases, the CONACYTS were an extension of the ideas and social groups that dominated the ancient academies; the only difference being that these groups, often isolated from their environments, suddenly found themselves with a certain degree of political support and funds.

This situation was recognized in its more formal consequences, even by the institutions that most strongly promoted the establishment of the CONACYTS: ". . . partly due to the composition of their management staffs and the initial impulse under which they were established, these Councils have neither proposed nor executed an S&T policy related to the socioeconomic needs of their countries . . . the Councils, and in general the national S&T system, have always been isolated from the other systems of the national reality, in particular that of the economic activities."(18)

The scientific elites, isolated from societal needs and perpetuating the operative characteristics of the academies, particularly the prestige patterns and the criteria for defining the priority sectors, were unable to find political allies for the support of their new bureaucratic situations.

Given the characteristics of the dependent development model followed by most of our countries, there was no reason why either the local or the foreign firms would become an autonomous source of demand for locally generated knowledge.(19) The multinational firms, using the advantage of their direct linkages with their parent firms, followed a predefined product cycle. The local firms, searching for foreign technology in order to survive in oligopolistic markets, had very little financial or human capacity to undertake, by themselves, substantial modifications in their products and processes. The governments paid little attention to the social needs in health, housing, and education that could be solved by means of research programs. In this context, who (which social forces) would support the programs proposed by the CONACYTS? Why should technology be a relevant variable in the overall development model as it was implemented in most Latin American countries during the 1960s and the 1970s?

Summarizing these ideas about the context in which most CONACYTS were born, we can say that:

- In most cases, they were the result of the historical coincidence between the "demonstration effect" of the scientific and technological boom in central countries, and the "planning fashion" that spread in Latin America after the Punta del Este Conference.

- Apart from some governments, which needed to use the CONACYTS as a source of political legitimation, and some scientific groups, which naturally supported the CONACYTS, there was no social group capable of (or interested in) supporting the action of the CONACYTS as a key element of their political project.

- The people, the criteria for choice of priority sectors, and, basically, the approach about the best ways of relating science and society were a continuum from the ideas of the academies.

THE CONCRETE EXPERIENCES

I shall make here a brief reference to the concrete case of the CONACYTS in four Latin American countries. With the single exception of Argentina, the cases have been taken from well-known authors who have analyzed the experiences of their own countries.

Mexico (20)

Since 1935, there have been certain institutions in Mexico in charge of coordinating and promoting R&D activities: The National Council for Higher Learning and Scientific Research (1935), became the Promoting and Coordinating Commission for Scientific Research in 1942, and in 1950, the National Institute for Scientific Research. In 1971, the National Council for Science and Technology – CONACYT – was created, with two main objectives: to aid the federal executive in an advisory capacity in activities in which CONACYT would not be directly responsible for execution; and auxiliary activities. These objectives included specific responsibilities, such as the working out of indicative S&T programs aimed at solving economic and social problems. But in general, the first type of functions assigned to the Council covered the full spectrum of R&D activities, as well as support activities (such as training and technology imports). However, the law that established the Council did not define the mechanisms available for the attainment of its ambitious objectives. For example, the law did not make any reference to all the technical activities which are the necessary link between research and production, for instance engineering design and technical assistance.

During its first two years of existence, the Council was seriously limited by the defects of its constitutive law and, therefore, without an adequate framework, it concentrated on channeling additional resources to international cooperation and to research institutions for training of human resources. In 1972, the Council developed a number of "indicative programs," whose purpose was to relate the research effort to the national development objectives. But these programs were suppressed after three months for purely administrative reasons.

This methodology of indicative programs was taken up again in 1973. And between 1973 and 1975, 13 programs were designed and implemented in areas such as health, housing, demography, use of marine resources, and meteorology. It is important to point out that the Council, in 1974, defined its role within the S&T system as that of "channeling additional resources ... having a relatively reduced participation because it must act over the system through the formulation of plans and programs, orienting and not centralizing the national efforts on S&T."(21) This marginal role, which CONACYT attributed to itself in regard to allocation of funds, is confirmed by the fact that, in 1975, the total budget of the Council represented only the 11.9 percent of the total R&D expenditure in Mexico.

As Nadal points out:

The relevant question with regard to this policy is whether it is possible to orient new research projects towards the national development objectives with a volume of additional resources which represents such a small percentage of the total R&D expenditure. This question is particularly important if we consider that the structure of the scientific system closely resembles what Alexander King calls the "pluralistic model," in which the decisions related to S&T are taken in an isolated manner, by different departments and agencies, without a general coordination.(22)

In order to overcome such limitations, the Council tried to strengthen its planning structure aiming at elaborating a global framework for the S&T policy, which would allow it to define medium and long-term priorities and criteria for fund allocation. For these purposes two new organisms were created within the Council: the interinstitutional commission on S&T; and the national commission for S&T planning. By mid-1976, the Council proposed a strategy for S&T development. The framework for this was the Indicative Plan on S&T (November 1975). This proposal has been qualified as "the most important effort for the establishment of a general framework for the S&T policy, not only in Mexico, but among the LDCs and can only be compared with the S&T plan of India."(23)

I will not go here through the main characteristics of the Mexican Plan, which have been brilliantly analyzed by Nadal in his book; however, I must point out that in early 1977, all this enormous effort was disregarded by the new authorities of the CONACYT, and the Plan rests only as an example of serious academic work.

Summarizing the Mexican experience, I can say that compared to its Latin American counterparts, the Mexican CONACYT received the most important economic support from its beginning. But given the lack of real social support, it went through a process of progressive bureaucratic complexity and change of strategies in an effort to fulfill its original objectives, given its marginal role in the definition of research policies, its limited resources, and the way they were allocated. It tried to widen its political influence through a comprehensive S&T plan, which failed precisely due to the lack of real social and political support.

The failure of this enormous effort seems to show that the political support given to the CONACYT was more formal than real and was conditioned to the changes in the bureaucratic structure of the government.

Argentina

The Argentinian CONACYT was created in 1969. Apart from the general, and well-known, objectives of planning and promotion, the law that established it assigned to the CONACYT the difficult task of coordination. Argentina has a very long tradition of scientific research based on a complex network of institutions with diffused responsibilities and various institutional affiliations. Among the preexisting institutions the CONACYT found we can mention:

- the Scientific Research Council (CONICET), an institution reporting directly to the President, whose mission was to "promote, orient, and coordinate the research activities in the field of pure and applied sciences." In 1972, the CONICET received 12.5 percent of the total national expenditure for S&T and had under its control 48 institutions, some of them through agreements with other public or private sponsors. In the "Program for Scientific Researchers," the CONICET had, at that time, 492 researchers who received allowances for full or parttime research. The greatest percentage of the support provided by the CONICET was concentrated on medical and biological sciences based on critiera of "international excellence." The CONICET was, and still is, the support of the "academic" science in Argentina.

- the National Institute for Agricultural and Cattle Raising Technology, created in 1956 as an institution for the promotion of research and transfer of knowledge to the primary sector. It is financed through a tax of 1.5 percent on the exports of the sector, and works through a complex network of research and extention stations located in almost all the provinces. It is responsible to the Secretariat for Agriculture and Cattle Raising.

- the National Institute for Industrial Technology (INTI), aimed at promoting the transfer of local technology, to industry. It is a part of the Secretariat of Industry, and, in 1971, was given the task of applying the law on technology transfer, through the creation of the Registry of Technology Transfer.

Among other institutions, we can mention: the National Atomic Energy Commission, founded in 1951, and reporting to the Presidency; the National Commission for Geoheliophisic Studies; and the institutes which are connected to the University which, in 1968, represented 74 percent of the human resources devoted to research activities.

The CONACYT did not receive funds for the promotion or support of S&T activities and thus, from the beginning, its activities were limited to promotion and coordination basically through the establishment of criteria and priorities for the "Science and Technology" item of the national budget.

From then on, the technical secretariat of the CONACYT tried to elaborate three national plans of Science and Technology, none of which received official approval.

In 1973, the CONACYT became the State Secretariat for Science and Technology at the Ministry of Education. This new situation allowed the CONACYT to receive some funds — though marginal within the overall expenditure on S&T — to be allocated as additional support to existing institutions through "National Programs" in various areas. But, at the same time, this new situation put the Council at the same administrative level with other research institutions which it was supposed to control and coordinate.

Between 1971 and 1973, the Council obtained from the Executive the approval of a law which allowed tax deductions to those private firms performing research and development activities up to a 100 percent of their expenditure on these activities. This experience was a complete failure because the technical staff of the Council had not developed a precise definition of the activities to be benefited with the deductions and, therefore, it became almost impossible to properly manage the requests.

It is worth mentioning that, during the period 1973-76, the Government defined a policy of explicit support for local S&T for basic elements of its overall development model. The CONACYT had no participation in the institutional structure thus created. Institutions such as the Corporation of Small and Medium Size Firms were created and located at the highest governmental level; and in all of them the promotion of the use of local technology was established as a priority task. The tax, which constituted the main income source for the INTI, was doubled, allowing the Institute to substantially increase its personnel and undertake new activities of crucial importance, such as its participation — as consultant — in the most important public works of the country.

The political model, based on the priority role attributed to the national bourgeoisie, gave an active political support to the creation and use of technology in accordance with the needs of the state and national firms. There was little room for the CONACYT in this model. In the cases where direct technological support was needed, INTA and INTI had the possibility to provide it. Where more research-based support was requested, either the universities or other institutions could undertake the programs without assistance from CONACYT.

With the government established in 1976, a radically different economic policy has been defined. It resorts to foreign technology as a basic element of its growth model, thus liberalizing the import of technology and closing, or reducing, the budget of most of the above mentioned institutions. The Secretariat of Science and Technology has continued playing a marginal role, currently reduced to the administration of some national programs, the effort to regionalize the scientific activities, and publication of a new survey of the scientific potential of the country. In these last two activities, it encounters serious conflicts with the recently created Ministry of Planning and other existing institutions.

Summarizing, I can hardly say that any substantial change has resulted in the quantity or quality of scientific and technological research in Argentina as a consequence of the existence of the CONACYT. Neither as a planning body, nor as a funding institution, has it exerted any influence on the creation and absorption of local knowledge. Once again, in the case of Argentina, no social sector has shown any interest in the activities of the CONACYT. In the case where a definite social group, the natonal bourgeoisie, expressed an interest in technology, such interest was channeled through other institutions such as INTI.

Brazil(24)

The history of scientific institutions in Brazil follows the same patterns as in many other Latin American countries. A number of institutions were created from the beginning of the century, but it was not until after World War II that a process of institutionalization began, with the State playing a leading role. In 1951, the National Research Council was created with the aim of "promoting the development of scientific and technological research in all the fields of knowledge." It was to act through fellowships, support for the acquisition of equipment, and exchange with foreign institutions.

During the government of Getulio Vargas, the State actively supported the S&T activities by founding a number of technological research institutions, in particular in industrial technology, mineral resources, and soil research. These institutions were aimed at supporting the needs of the import substitution process that Brazil began after World War II.

All through the 1960s, Brazil developed a process of progressive internationalization and oliogopolization of its economy with the priority role played by the multinational firms and, in some sectors, the State. The opening of the economy determined the massive use of foreign technology and a restriction on support to the already existing infrastructure. Vera Candido Pereira affirms that "the 15 years that run between 1955 and 1970, characterize a dark period for S&T in Brazil, given the lack of any structural need for their development."(25)

With the beginning of the 1970s, the State expressed a definite interest in support for S&T activities which is reflected not only in the support for the existing institutions and the creation of new ones, but in a very complete exercise of S&T planning. Changing its previous statements, the government explicitly talked of "substitution of technology imports," and created a complex institutional network with the allocation of the most important amount of resources of any kind in the history of Brazil for this purpose. This experience has some features which are worth analyzing in detail.

The National Research Council was restructured and received new functions, as: 1) "to assist the Ministry of Planning in coordinating the execution and formation of the Basic Plan for S&T Development, as well as in analyzing sectoral plans and programs of science and technology; 2) to stimulate the execution of programs and projects of interchange and transfer of technology between public and private entities, national, foreign, or international, etc."(26) The real promotion and financial support of S&T activities rest on a complex of institutions, in particular the FUNTEC – National Fund for Technological Development – and FINEP –Financier of Studies and Projects – both at the National Development Bank. These institutions provide the funds for both private and public sectors, playing a role of integration between the proposals of the Plan and the priorities established by the industrial development authorities. It is worth mentioning, too, the "Groups for linkage with industry" (CCNAI), whose aim is to integrate the specific requests of capital goods from the state enterprises, with the production of the machinery sector, and in which the FINEP has a leading role.

The second interesting fact is that the Brazilian Plans establish as priority sectors those with more sophisticated technology: nuclear, energy, space, economic infrastructure, etc. Many authors question whether these areas correspond to the real needs of the productive sectors of the country, especially those of small and medium size firms. It is clear, however, that the choice of these sectors shows one of the reasons for this sudden interest of Brazil in S&T; it plays in the Brazilian model an ideopolitical function, aimed at supporting the objective of "putting Brazil, in the time of one generation, among the great world powers," as stated by the First National Development Plan.

The increasing restrictions on the foreign sector have acted, too, as an impetus for the development of technologies aimed at finding new export markets and at accelerating the substitution of imported raw materials and capital goods.

Summarizing, in the Brazilian case there seems to have been a sharp change in the attitude of the government towards S&T activities during the last few years. I have been unable to find in the literature an evaluation of the results obtained through the application of the First (1973-74) and Second (1975-77) Basic Plans for S&T Development. The nature of the sectors given priority in both Plans, as well as the structural restrictions posed by the development model in force in Brazil, will require a number of years before the first results can be realized.

For our specific interest, I must mention the fact that the CNP seems to play in the overall model a complementary role – that of planning and advising the executive – while the concrete execution of policies rests on a number of institutions not being controlled by the Council.

Peru (27)

The National Research Council was created in Peru in 1968, in one of the first legislative acts of the Revolutionary Government of the Armed Forces. The creation of the Council was the result of the efforts of a number of local scientists, and the influence of some international organizations – most notably the OAS. During its first years, the Council undertook a number of studies on resources for S&T and on transfer of technology, which gave a detailed picture of the scientific potential of Peru at that time. But these were the only remarkable activities of the Council during these years. As Sagasti points out: ". . . the Council was structurally unable to perform the functions assigned to it by law, which were defined at a time when the confusion between science policy and technology was the rule. . ."(28) The Council had been wasting its political possibilities trying to formulate policies for a number of institutions it did not control, or developing ambitious programs for which it did not have enough resources.

The General Law of Industries, in its articles 14 and 15, created the ITINTEC and also the industrial technology research fund, formed by allocating two percent of the net income (before tax) of industrial enterprises. In addition to the ITINTEC system, the general laws of mining, fishing, and telecommunications contain similar clauses that direct the same two percent of each sector's net income for technological research. This pattern of financing assured a stable source of funds for the institutions thus created, and represented a new and very effective way of dealing with the S&T problem in Latin America.

The ITINTEC system "can be considered as a multiple function organization which operates several policy instruments to develop technological capabilities in Peruvian industry."(29) The organic law of ITINTEC and other sectoral institutions assigned them the functions of promoting, supervising, and carrying out technological research; preparing technical norms and standards; improving quality control; and additional activities such as dissemination of technical information and training.

I will not analyze here in detail the particular operative patterns of the ITINTEC and the other sectoral institutes, which can be found in the excellent paper by Sagasti which serves as the basis for this description. However, more than 200 projects generated in the industrial sector itself were approved and executed and ITINTEC has contracted with local research institutions for an important number of projects aimed at solving definite needs of industry, which shows the success of this unique system.

It is a success that can be explained by the following facts:

- The system of sectoral technological institutes is born not as an isolated administrative decision, but as a part of a whole development project based primarily on the strengthening of the national firms, either private or state owned.

- The system receives not only objectives, but concrete institutional and financial support to fulfill them.

- The institutes cover the whole spectrum of the activities related to technological development in their sectors; and are used as a linkage between the needs of the industries and the solution of their problems with the research capabilities of the country.

SUMMARY AND CONCLUSIONS

In this chapter, I have tried to analyze the problems of the Science and Technology Councils – CONACYTS – without resorting to the typical statistical or descriptive approaches.

There are two central ideas in the scheme I use. The first is that the CONACYTS are more an invention of the theorists than the natural result of the development of productive forces in our countries. Theoretical schemes were built up, copying the experiences of some central countries, assuming that the reality would safely suit them. The CONACYTS should play a catalytic role located at the highest bureaucratic level, and be endowed with a great formal power.

The fact that the environment could be hostile or indifferent to the products and proposals of the S&T system was not explicitly considered, assuming, instead, that it was only a matter of the embryonic social situation of our countries which was causing delays.

The State, a central element of the whole process, has been considered as a merely administrative institution which should also mature towards a greater understanding of the potential role of S&T. In order to make clearer these ideas, I have described the purest approaches that have been used.

Through a close examination of the history of these processes, I have tried to find which were the main limitations experienced in the implementation of these proposals. My idea, in this respect, is that the main obstacle is precisely the hostility or indifference expressed in the lack of explicit social support. Without any structural need for locally generated S&T in our countries, given the features of the dependent model in force, and with the potentially interested social actors (national bourgeoisie) limited in their possibilities for political expression, who would fight for an indigenous science and technology?

Without social support, the most ambitious formal efforts (CONACYTS and Plans) have no chance for survival. Such is the pathetic case of the Mexican Plan and the Argentinian CONACYT. With political support, the CONACYT can be one more component in a more complex apparatus; but given the characteristics of our productive structures, it can never be the institution.

The policy conclusions derived from this analysis are not necessarily pessimistic or fatalistic.

A first conclusion should refer to the need of a complete revision of the theoretical approach used to treat the problem. The new approach must be essentially historical and conjunctural, designed for each time and space and taking into account, as a central element, the role of the potential allies and enemies that different policies may find.

If the environment is hostile, it will be necessary to define the more adequate institutions for such a situation. For that purpose, we also need an approach of here and now. Past experience shows that we will probably have to leave aside the fascination with the "big science" represented in the CONACYTS, and search for more direct and efficient instruments. Among these, I think that the sectoral technological institutes can play a central role. In a situation of scarce resources, with the need to concentrate and combine various functions, as well as to search for the most direct possible contact with the entrepreneur, these sectoral institutes have proven to be very efficient.

The message for the United Nations Conference is straightforward: a substantial part of the critical and out-of-context proposals I have referred to have originated, and still are being generated, from the well-known international experts of the United Nation Agencies. A theoretical revision, therefore, should start with the ideas that lie behind such proposals.

The Conference may propose, once again, that the only problem is lack of resources devoted to the development of S&T and that, in consequence, all that is needed is more money, more people, more institutional resources; in other words, to deepen the linear development of the model. If this is so, and the recommendations and plans that stem from it are of this sort, we will probably assist in the production of a new conference, in ten years time, devoted to the same kind of discussion of the same kind of problems.

NOTES

(1) See for example C.V. Vaitsos, "The Process of Commercialization of Technology in the Andean Pact" in H. Radice, ed., International Firms and Modern Imperialism (Harmondsworth: Penguin, 1975).

(2) See the classic of this line: F. Sagasti, "A Systems Approach to S&T Policy Making and Planning" (OAS, 1976).

(3) Ibid., p. 3.

(4) Ibid.

(5) Ibid., p. 13.

(6) Ibid., p. 80.

(7) See Oscar Oszlak, "El Area CT del Estado Argentino," STPI Project, Argentina.

(8) Ibid., p. 17.

(9) Eduardo Amadeo and Liliana Acero, "A Bibliographical Survey on Industrial Research Institutes; Suggestions for Further Research," unpublished consultancy study for the IDRC, 1977, p. 3.

(10) See, for example, the various papers by Alberto Araoz on this matter; or Luis J. Jaramillo, "El Papel de las Compras del Sector Publico" (Bogota, mimeo, 1975).

(11) Op. cit., p. 15.

(12) See, for example, Ruth Sautu and Catalina Wainerman, "El Empresario y la Innovacion," OAS, 1972.

(13) FINEP, "Difusao de Innovacoes na Industria Brasilera, Tres Estudos de Caso" (Rio de Janeiro: IPEA, 1976).

(14) Ibid., p. 5.

(15) Vera Maria Candido Pereira, "Reflexoes sobre Estado, Ciencia e Tecnologia no Brasil," mimeo, FINEP, Rio de Janeiro, 1976.

(16) Ibid., p. 18.

(17) See Alberto Sanchez Crespo, "Esbozo del Dessarrollo Industrial de A.L. v de sus Principales Implicaciones Sobre el Sistema Cientifico y Technologico," OAS, 1972.

(18) See UNESCO, "La politica cientifica en A.L. "en Estudios y Documentos de politica cientifica," no. 37, p. 14.

(19) See Sanchez Crespo, op. cit.

(20) The analysis of the Mexican case is based on the work done by the Mexican group of the STPI Project, and the excellent book by Alejandro Nadal, Instrumentos de Politica Cientifica y Tecnogica en Mexico, (Mexico City: El Colegio de Mexico, 1977).

(21) See Gerardo Bueno "Atribuciones, estructura y programas del CONACYT," CONACYT, Mexico, 1974.

(22) Nadal, op. cit., p. 25.

(23) Ibid., p. 28.

(24) Based on Eduardo Augusto de Almeida Guimaraes and Ercila Ford, Science and Technology in Brazilian Development Plans, FINEP, 1975; and Vera Matia Candido Pereira, op. cit.

(25) Pereira, op. cit., p. 42.

(26) From "Legislation concerning the CNP," Rio de Janeiro, Brazil, 1975.

(27) Based on Francisco Sagasti, "A framework for the implementation of technology policies: a case study of ITINTEC in Peru."

(28) Ibid., p. 34.

(29) Ibid., p. 40.

12 Science and Technology Planning in LDCs
Miguel S. Wionczek

The recent information about the working of the S&T systems* in some more advanced underdeveloped countries, such as Argentina, Brazil, Egypt, India, or Mexico, strongly suggests that their growth faces many formidable internal obstacles. In a recent resolution adopted by the Preparatory Committee for the UNCSTD (A/Conf.81/PC/L.12), these obstacles were identified as related, among others, to policies and priorities for science and technology, infrastructure, systems for education and training, availability of entrepreneurs and managerial skills, human and financial resources, S&T information systems, and technological extension services. The fact that, in spite of increased financial support offered S&T by the state and international organizations, and in spite of local and regional S&T planning attempts, all these obstacles persist strongly suggests that the task of building up the domestic S&T capacity in the LDCs is much more difficult and complicated than it has been assumed in many quarters both within the United Nations and at the national level.

Practically all major LDCs continue to depend to an overwhelming degree on the knowledge and the technical know-how produced in the advanced countries. In most instances, domestic R&D activities in the LDCs consist of imitative quasi-research in pure and applied sciences, with very little progress in technological innovation, including fields in which serious and competent indigenous R&D has been badly needed for quite some time, for the very simple reason that many economic and social problems arising in the context of overall underdevelopment are quite different from those presently facing the high-income countries, whether endowed with market-oriented or centrally-planned economies.

Whatever R&D efforts are made in the LDCs locally, sometimes at

---

*Defined here as the total of scientific, technological and industrial development research centers and the tangible and intangible scientific and technological infrastructure.

considerable financial and social cost, they are often largely wasted due to the absence of permanent links between S&T activities and local educational and productive systems, which leads to a lack of correspondence between the demand for, and the supply of, internally produced knowledge and technical know-how.

LDC scientists and technologists participate only marginally in higher education activities, even in cases where the few existing R&D centers are physically located at the universities. The unsatisfactory conditions of general technical education result, in turn, in great shortages of the kind of technical staff needed to support R&D. These shortages of middle-level human resources for S&T, together with the bureaucratization of the public sector and the inadequacy of higher education institutions, depress further the generally low productivity of local scientific and technological communities. Moreover, the absence, or the weakness, of S&T diffusion mechanisms at all educational levels perpetuates the isolation of S&T from the society, and impedes the extension of the scientific and technological culture outside of the small community of elites directly engaged in R&D.

The demand of the productive system for technical know-how and innovations is satisfied mainly from abroad, even in the most advanced LDCs. Productive units, whether private or public, and whether national or foreign-owned, show clear preference for foreign-originated technology. The productive system assumes that local scientists and technologists are unable to produce either useful knowledge or the process and product technology which could progressively supplant S&T imports. Moreover, from the viewpoint of risk aversion, imports are more convenient than unproven local know-how.

Not only are the S&T systems in the LDCs disregarded, due to quantity and quality of human resources required, but financial resources available domestically for R&D and S&T infrastructure continue to be largely inadequate, both in absolute and in relative terms, when the per capita income differences between the LDCs and the developed countries are considered.

The functional distribution of R&D expenditures in LDCs is deficient and inhibits S&T expansion. While up to 75 percent of financial resources, available mainly from the state, is spent on salaries and wages of R&D personnel, funds for both hardware and software, indispensable for serious and relevant S&T efforts, are extremely scarce.

While most R&D institutions – whether located at the universities or in the public sector – display critical lack of researchers, their proliferation is the order of the day because of the false and superficial self-prestige of their directors. In consequence, very few scientists and technologists dedicate full time to R&D. Even worse, in the face of the general absence of R&D managerial skills, the best brains become rapidly absorbed in S&T administrative tasks, and readily move into other bureaucratic activities.

The S&T development is highly unbalanced both sectorally and by disciplines, with a consequent neglect of very important R&D areas. This reflects two facts: first, the S&T centers and institutional network in the LDCs is the result of haphazard and uncoordinated individual initiatives; and second, most decisions concerning research areas and priorities are taken by the foreign-trained S&T personnel, whose preferences, in spite of their operative competence, lead them to engage in the kind of research in vogue in more advanced countries.

While in most LDCs the supply of so-called "pure" researchers is larger, for historical reasons, than that of staff interested in applied research and technological innovation, financial resources for basic research are extremely scarce. Because of lack of interest in R&D on the part of the private sector, neither are there sufficient funds for applied research nor for technological development. These activities are concentrated in but a few sectors of the more advanced LDCs, in which the presence of the state and the public sector as the producer of strategic industrial goods and social services is particularly strong. Thus, for example, in Mexico in the past 10 years petroleum and energy, commercial agriculture, and medicine and health were absorbing about one half of the total financial resources available for S&T. On the other hand, all LDCs – including Mexico – have neglected R&D in the areas of key importance for the equilibrated and equitable socioeconomic development such as subsistence agriculture, nonrenewable (except petroleum) and renewable resources, capital goods manufacturing, transport and communications, or urban development and housing.

The expansion of S&T activities in the LDCs is hampered also by a host of other factors, arising from the fact that whatever weak and sporadic attempts are made to allocate more financial and human resources for S&T, these allocations are neither supported by a long-term S&T strategy, nor linked with general socioeconomic planning. The LDC political elites do not, as a rule, relate S&T activities with the socioeconomic development needs. Ignorance of this relationship is even greater among domestic entrepreneurial groups.

Contrary to evidence provided by the advanced countries, in most LDCs, S&T is looked upon as just another sector which – it is thought – should receive more support than in the past just because the world lives in some sort of a dimly perceived (by the LDCs) technological revolution. This primitive perception of the importance of S&T translates itself into the appearance, from time to time, of S&T crash programs, which are supposed to close the S&T gap between the LDCs and the DCs. The crash programs amount to little more than sending potential researchers abroad for training, and/or to small increases in the budgets for existing R&D institutions. As a rule, however, foreign training of human resources is not coordinated with increased allocation of funds for R&D; neither are efforts made to integrate these crash programs into the overall S&T priorities and objectives. Thus, results are disappointing, particularly from the vewpoint of the LDC politicians, who rarely understand the long-term nature of S&T advancement. In the absence of anticipated quick, tangible results, the brief periods of official S&T support are followed by abandonment of crash programs in that field, and anti-intellectual recriminations against local S&T elites.

Moreover, practically nobody in the LDCs realizes, as yet, that S&T

consists of a broad spectrum of activities which affect all spheres of national life, and that, in turn, S&T advancement is affected by all sorts of policies in force. Consequently, isolated actions, such as the allocation of funds for improving the quality of human resource engaged in R&D and the expansion of R&D facilities, may be completely nullified by non-S&T policies which continue directing the weak domestic demand for knowledge and technical know-how toward external sources.

This happens even in those more advanced LDCs where attempts have lately been made to engage in S&T planning, because most non-S&T policies in force had been designed long before the recent emergence of half-hearted recognition of the role of S&T in the LDC socioeconomic development. No attempts have been made to integrate S&T policy actions with economic and social policies, and many traditional (particularly economic) police instruments have a perverse impact upon the objective of creating an indigenous S&T capacity. The results of the uncoordinated S&T policy actions are highly disappointing. While the conflict between S&T and other policies is particularly visible in the field of LDC industrialization, it can be discerned also in other important sectors, such as agriculture, health services, or urban development.

This brief survey of major internal obstacles facing the S&T advancement in the LDCs permits us to define an overall framework for S&T planning — a framework which represents the necessary, but not sufficient, condition to create, in the longer term, some sort of autonomous S&T capacity:

1. S&T planning must be incorporated into general long-term socioeconomic planning.

2. The major objective must be the establishment of links between domestic production of knowledge and technological know-how on the one hand, and the economic, productive, and political systems on the other.

3. Very high priority must be given to the task of diffusing scientific and technological culture throughout the society or, using economic terms, to the objective of creating some degree of preference by local S&T consumers for domestically-produced knowledge and technical know-how.

4. It must go beyond the sphere of R&D activities as defined in advanced countries, where not only the S&T infrastructure is strong and diversified, but permanent links between R&D activities and the educative, productive, and political systems established themselves over a long time.

5. It must charge itself with the building up of the S&T infrastructure in the broadest sense, consisting not only of human resources (for managerial, R&D, and research-support activities)

and the network of R&D centers, but also of such R&D support services as diffusion and information gathering mechanisms, recollection and processing of statistics, data computation, engineering and consulting firms, production and maintenance of R&D equipment and instruments, and technical standards.

6. Some priorities must be established for very broad R&D areas, taking into consideration not only the long-term socioeconomic development objectives, but the present and potential availability of human resources; S&T planning is meaningless if it is limited to an elaboration of a consolidated list of possible R&D projects which might be undertaken if additional financial resources were available.

7. It should, under no circumstances, be led by the mirage of autarchy into the production of knowledge and technical know-how, in spite of the political appeal of such an objective in some LDCs. It should be based on the premise that the externally-available knowledge and technology can be, in the medium run, only supplemented with local R&D production, and that the major role of local R&D is to increase gradually the degree of national self-reliance with respect to the decisions of the application of knowledge and techical know-how, whatever its origin may be.

8. Except when the externally-produced knowledge and technology is unavailable or – albeit available – its application conflicts with the major national socioeconomic development objectives, S&T planning must give the highest priority to the building of domestic capacity to absorb and adapt S&T produced externally, as a precondition to indigenous production of knowledge and technical know-how.

9. Because the building up of the indigenous capacity to absorb, to adapt, and eventually to produce knowledge and technical know-how can be successful only if accompanied by the emergence of domestic demand for it, S&T planning must provide for instruments regulating technology imports – instruments that could be applied not only to technology transfer transactions, but to industrial property and direct foreign investment as well.

10. Finally, S&T planning must make a distinction between science planning and technology planning, even though it is almost impossible to establish clear limits between scientific and technological activities.

The contemporary literature on S&T is full of idle speculation about the relationship between science and technology. The little that is known about this relationship is that it is not linear, and that modern S&T systems started, historically, from many different points of the continuum. In the early nineteenth century, there were the countries which excelled in science (France), along with the countries which had relatively little science but a lot of technological progress (Great Britain). There were also the countries which lacked both (the United States), and where S&T started through technological experimentation. One and a half centuries later all these countries are major S&T powers.

It is often maintained in the LDCs that science advancement is a luxury which they do not need, or can hardly afford. Concomittantly, voices are also heard in the LDCs to the effect that the only science which should be promoted is that which will result in very rapid technological applications of its findings and discoveries. Both attitudes, which are not only incorrect but socially damaging, originate with pragmatic ignoramuses. The LDCs need both good science and the relevant technical know-how for reasons that go beyond the common sense proposition that good science is helpful to the production of technology. Science has other important functions besides supporting the expansion of technical know-how. The most important function, perhaps, is that of providing a base for a general scientific culture, badly needed for an increased degree of overall rationality in LDC societies faced with the most complicated and pressing social and economic development problems. While scientific attitudes are not expected to substitute for ideologies, the use of ideologies, in the absence of control mechanisms provided by science and rationality, has led in the past, and will lead in the future, to major political and social disasters.

The LDCs must promote science for other reasons as well. Scientific advancement is an important source of national satisfaction and prestige. It offers permanent linkages to the outside world, and provides socially useful occupations to clusters of scientists which, however tiny, represent the kernel of S&T elites in the LDCs. Disregard for, or the abandonment of, science by the LDCs would clearly result in the permanent emigration of first class human assets, which would represent a very considerable loss to the society. Such emigration would take place even if an LDC gave the highest priority to technological advancement, because psychological characteristics of researchers, social determinants, and the values involved in the production of knowledge are quite different from those involved in the production of technical know-how. These differences explain why first-class scientists are very rarely first-class technologists, and vice-versa, although always there will be exceptions to this general rule. Together with the differing nature of scientific and technological endeavors and functional differences between science and technology, social and psychological aspects of S&T explain why scientific planning offers a different sort of challenge, and calls for approaches and methodologies different from those involved in technological planning.

It is sometimes maintained that, because of the unpredictability of scientific discovery, science does not lend itself to any planning. The statement is true only in cases where planning attempts to impose upon science short-run criteria aimed at assuring practical application of scientific progress. Since such criteria contradict the very nature of scientific discovery and advancement, the planning of science should not even attempt to elaborate them.

The impossibility of establishing pragmatic criteria for science does not invalidate, however, the need for, and possibility of, planning for science in the LDCs. Science can and should be planned: first, because of the scarcity of human resources for S&T; and second, because of the infinite choices facing scientific endeavor and speculation. The major objectives of science planning are – in the broadest terms – threefold: 1) to increase scientific productivity and assure excellence, i.e. to produce good and relevant science at the lowest social cost; 2) to create conditions for an interdisciplinary approach to scientific problems, and 3) to advance the frontiers of science in such a way that scientific advancement may assist, directly or indirectly, applied research and technological development, neglected by S&T effort in the advanced societies.

Both S&T planners and scientists in the LDCs must be aware, however, that the presence or the absence of possible linkages between pure research and its subsequent application cannot be determined <u>ex ante</u> and that, furthermore, the application of science to the solution of nonscientific problems cannot become the sole or principal criterion for scientific support. History abounds in its examples of scientific discoveries that were followed by their nonscientific application only after the delay of many generations. Other great scientific discoveries have never been translated directly into technological progress.

Contrary to some misconceptions, judicious science planning does not involve the abdication of scientific freedom by researchers. The task of S&T planners, all of whom must have some experience with the nature of scientific advancement, is limited to defining, in very general terms, those wide areas or categories of disciplines in which scientific progress seems more likely than in others. They must take into consideration the availability of highly competent human resources and infrastructure, and maintain some sort of balance between the accumulated knowledge and the perceived scope of ignorance. These elements must be carefully weighed – again, in very broad terms – against alternative needs for scientific progress in an underdeveloped society. The complicated process of arriving at general priorities in respect to support of scientific activities – in the LDCs and elsewhere – should never fall into the trap of absolutist rigidity. Neither should it degenerate into decisions about the social relevance, or irrelevance, of specific research projects.

While it seems easy to arrive at an abstract conclusion that research in biology may be more relevant for an LDC than the pursuit of astronomy or topology, translation of such priorities into either support, or the withdrawal of support, for specific disciplines or – even worse – for specific research projects, clearly courts a disaster. The most S&T planners in the LDCs can expect and hope for is that the combination of human resources available for, and willing to engage in, scientific activities and financial resources will produce good science, and that the support of science by the state will increase the socialization of scientific elites, i.e., improve their grasp of the social function of science.

Achievement of an increased degree of social consciousness on the part of participants in local science efforts in the LDCs should be considered a great feat by itself. Scientists in the LDCs are more asocial than their counterparts from the advanced societies, not only because they share certain non-social values and are the product – as anyone else – of the faulty LDC educational systems, and victims of the overspecialization which characterizes much modern science, but also because they are often rejected by these same underdeveloped societies.

Armed with very broad guidelines in respect to preferred areas of scientific endeavor, established within a flexible framework, and engaged in consistent long-term support for a scientific infrastructure, and in sustained efforts to diffuse science and a scientific culture, the S&T planners in the LDCs can expect positive responses from local scientific communities. LDC scientists are clearly able to make a distinction between outside interference with scientific freedom, in the name of real or ill-conceived social needs, and the genuine support for science, viewed as an important segment of the S&T continuum.

While science planning cannot be left to scientific communities alone, neither can it be engaged in without their direct participation; planning and programming of research must be left to the responsibility of scientists and their institutions. The role of the overall S&T planners as advisors at these levels is still very considerable. They can help in designing long-term research programs, and act as innovators in the institutional field. They should also devise and implement long-term science support mechanisms, with respect to the supply of both the research hardware and software, and to propose measures that would make the daily operation of scientific institutes and individual researchers more secure – financially and otherwise. The reasonable working of such mechanisms and measures does not depend on the provision of funds alone, but also on the elimination or alleviation of manifold bureaucratic obstacles present in abundance in any underdeveloped society and due, in the case of science activities, also to general ignorance about the importance of S&T vis-a-vis other activities.

In brief, science in the LDCs should be underplanned rather than overplanned. Furthermore, the planning should be directed to the outer fringes of the scientific endeavor and to its infrastructure and not to the substance of scientific research itself. Clearly, planners should avoid falling into the other extreme of catering indiscriminately to those segments of LDC scientific communities which just demand more money from the state, but are unwilling to accept any social compromise, claiming to be citizens of the alleged 'free republic of science.'

Planning of applied research and technological development in the LDC is another story. It just cannot be done without entering into the substantive problems of applied research and technological development. This exercise, in turn, is practically impossible in the absence of a detailed diagnosis of the state of art with respect to the technical know-how available from different sources, foreign and local; of the degree, if any, of the indigenous technological innovation; of the channels for technology transmission; and of policy instruments of all sort which affect imports, in any important degree; and the local production of technical know-how, as well as the global and sectoral demand for technology. Even when all such information is collected and analyzed, planners will still be unable to establish criteria and fix priorities for technology imports, their local adaptation, and the production of technical know-how, unless they can relate these criteria and priorities to a longer-term socioeconomic development model to be pursued. In other words, unless national policy makers at the highest level have a clear idea about what kind of society they want to construct, no coherent technological policy is possible. In such a case, it will be the type of technology imported, largely on grounds of its private profitability, that will decide the future shape of the society.

Ideally, technological planning in the LDCs should be incorporated into overall socioeconomic long-term planning. In most LDCs, formal socioeconomic planning more resembles science-fiction, or a ritual rhetoric, than a real national exercise. In such a situation, relating technology planning to what some social and political writers call a national project of the future society may represent the second-best, but realistic, solution, because it may be the only way to permit a first approximation of the overall long-term technological needs of the society in question.

The next task of the planners would be to define and delineate those economic and social sectors whose different functional characteristics decide the kind of specific technologies needed. In the case of S&T planning in Mexico, long debates that took place on that subject involved public policy makers and the private sector, producers and users of the technical know-how, and leading educators. The decision originally arrived at by negotiation was that there was a need to make distinctions among technological policies and priorities with respect to the following sectors: agriculture and forestry; fisheries; manufacturing industry; mining; energy; transport and communication; urban development, including the construction industry and housing; medicine and health; and education. This division of technology by sectors was found later to be in need of refinement because it did not take into account certain strategic economic activities that could only fall into the category of a sector with difficulty; it overlooked the important intrasector differences present in some sectors; and it forgot some sectors which were not productive in the limited economic or social sense. Consequently, the list of technological sectors was amplified by 1) adding foodstuff production which covers the whole range of activities from agriculture to distribution of processed foodstuffs; 2) subdividing the manufacturing activities into consumer nondurables, intermediate goods, and consumer durables, considered jointly with capital goods; 3) subdividing agriculture into commercial and subsistence agriculture, and 4) establishing a new broad category of activities for which there is a need for both applied research and the technical know-how to take care of such problems as ecology, renewable resources, and natural phenomena.

The elaboration of a similar technological map of the economy and the society is a prerequisite for the following planning stage that involves the sectoral diagnosis of the technological demand and the sources of supply. Such an exercise, for which the best national experts (including scientists) must be mobilized, implies the availability of a considerable body of knowledge on the general state and direction of the world-wide technological advancement in individual sectors or subsectors. Detailed information must be collected about technical imports (also by sectors) and about the sectoral state of art in local R&D. As no single national public agency and no single local R&D institution has the information needed at that planning stage, it can be successful only with the broadest participation by national experts in each field, preferably on an individual, nonpolitical basis.

This stage of technological planning involves also the establishment of a permanent S&T statistical office, able to process detailed information about the national S&T system in terms of its institutional structure, human resources, and R&D programs and projects underway or planned.   Such a statistical S&T office must work in close cooperation with national agencies in charge of regulation of technology imports, industrial property (patents and trademarks) policy, and the control of direct foreign investment. The latter is a major channel for transmission of foreign-originated technical know-how.

The following stage corresponds to the translation of the sectoral diagnosis of the supply and demand of technology into general R&D guidelines, for a period of at least 10 years, to be revised periodically in the light of technological advancements both within and without the country, and in the light of possible changes or adjustments of domestic socioeconomic policies. The elaboration of such guidelines would involve the full presence of representatives of all major local R&D institutes, and of major technology consumers in both the public and private sectors. Eventually, the ensuing substantive guidelines – which cannot be considered iron-clad priorities – will indicate the possible R&D areas of broad subjects. Such guidelines will have to be treated in a very flexible manner, and with the understanding that the reasons for the exclusion of some areas or subjects from such sectoral lists are threefold: either these R&D areas can be taken care of by technology imports, or they are not highly relevant to the economy and the society in question, or no human resources are available domestically for their meaningful pursuit. The broad scope of substantive guidelines, which must be indicative and not mandatory, offers a great advantage for the planning implementation stage. It will permit R&D institutes to work out programs and projects tailored to both their human resources endowment and their financial possibilities. It will also permit them to adopt some general criteria with respect to the training of new researchers and the creation of an institutional infrastructure for the more distant future.

The final task of S&T planners, common to both science and technology, is that of estimating the domestic S&T system capacity to absorb additional financial resources in the light of 1) the possible supply of new human resources, and 2) the expenditure planned for the expansion of S&T physical and institutional infrastructure. This exercise, which will make use of all information – both quantitative and qualitative – collected during the process of planning, belongs to the planners themselves. This exercise, that will make it possible to set long-term financial targets for S&T support, must be done simultaneously from two ends: it must consider, at the same time, the potential availability of total financial resources for S&T; and also the probable costs of 1) maintaining the existing S&T system, 2) training and employing new S&T personnel, and 3) improving and expanding the S&T infrastructure over the planning period. Attempts to set the broad, overall, national targets for S&T expenditures – in terms of the proportion of the GNP or the per capita income – which do not consider the S&T system's financial and human resources absorption capacity are useless.

At this point, the S&T planning, in a formal sense, will be accomplished. It will clearly be of little use, however, unless it is followed by an implementation stage, whose description and analysis lies beyond the scope of this brief chapter. It is worth mentioning, however, that the implementation cannot be limited to the mere elaboration of a sequence of S&T sectoral and institutional research programs and projects for the duration of the Plan. It will have to involve many very important political and substantive decisions with respect to such issues as the shape and the power of the permanent S&T plan implementation agency and its position within the Executive Branch, legal changes needed to eliminate conflict between the S&T policy instruments and non-S&T policies, and concrete ways and means of expanding and strengthening S&T infrastructure. The scope and the importance of the implementation of the S&T planning strongly suggests that institutional arrangements created originally for planning purposes should not be dismantled after the elaboration of a plan. These arrangements will have to be redesigned, however, in the light of the experiences acquired during the planning process.

Considering the UNCTAD interest in the subject of S&T planning and advancement, it is suggested that the matter of the relationship between science planning and technology planning be studied further with a view towards presenting a relevant paper to the UNCSTD. Few such papers, if any, can be expected from other quarters. It seems advisable to mention that such a paper should not advocate the divorce of scientific planning from technology planning in the LDCs. Whereas the present division of labor in the United Nations family can make such a proposal attractive on the grounds that science and technology deal with "different things," such a divorce, if implemented, would be very harmful. What is needed is the recognition by the United Nations' family that while the UNESCO approach to S&T problems, putting all the stress on science cum culture, is not correct, neither would it be correct to put all the bets on the LDC technological advancement and forget science.

13 Science and Technology Policies in Latin America— Against a Holistic Approach

Joseph B. Hodara

A holistic approach characterizes the design of many science and technology policies in Latin America. The prevailing tendency is to regard science and technology (S&T) as inseparable components of a unified process of knowledge production.(1) One sign of this tendency may be found in the very name of most national institutions in charge of S&T in countries such as Mexico, Barbados, Dominican Republic, Venezuela, and Argentina. Both components appear therein as subjects of public and governmental concern. Moreover, the formal range of functions assigned to these bodies, as well as the substantial considerations which legitimate them, reflect a comprehensive view with respect to the factors, channels, expressions, and consequences of scientific and technological progress.

For example, referring to the Dominican Republic, Herdia and Vega say that the "Unit for Science and Technology (UNICYT) has a twofold function: to outline national priorities and to suggest proposals for the development of science and technology . . . UNICYT seeks to coordinate the exercise of science and technology with the established national priorities."(2) And making their point clearer they add: ". . . a comprehensive science policy embraces science and technology activity in all its aspects, from the determination and dissemination of knowledge to its implementation in the production and distribution of goods and services."(3)

A recent study made in a Central American country moves in this same direction. It states that "the science and technology system of Nicaragua presents some weaknesses that should be corrected."(4) Yet the analysis made therein reveals that these weaknesses are so fundamental that one hardly may speak of a system.

This holistic stand is also present in the current and continuing debates about the cause of the scientific and technological lag. According to them, S&T coalesce into one single factor which causes – and explains as well – social and economic underdevelopment. Thus, a document produced by ICAITI (Guatemala) suggests that "the constraints upon the utilization and diffusion of techniques flow . . . from the insufficient development of the science and technology system, which results in the external dependence and the backwardness of wide segments of Central American economy and society."(5)

In sum, if S&T make up a "single variable" they should be treated simultaneously.(6) A comprehensive management of S&T policies is needed. From this perspective, the specific patterns and rhythms which characterize different fields, disciplines, and productive sectors are to be disregarded.The recommended approach will take care of them all.

Notice that we fairly distinguish between a synthesis of S&T, and an all-embracing treatment. Indeed, the first is proper; the second, we shall argue, is counterproductive.

This holistic approach has antecedents – and perhaps inspiration – in work done by several international and regional entities. For example, key United Nations documents, which discuss science and technology issues, feature a remarkable comprehensiveness.(7) This is, of course, justifiable at least in part, considering the diversity of problems and audiences to which those documents address themselves. Likewise, most papers produced by the Organization of American States (OAS) in the early 1970s exhibit this same trait.

In the context of S&T policies in Latin America, holism involves at least three assumptions. First, that S&T, being parts of a continuum, reinforce each other. That is, actions undertaken at one point (i.e. basic research) would directly have a favorable impact at another point, i.e. marketable innovations). Having established a linear relationship between them, public policies have to sustain a felicitous chain reaction among the different stages and segments of knowledge formation and usage.

Second, holism spells centralization, and centralization in politically totalist societies(8) implies that the government represents the entity called to promote exclusively S&T endeavors – not only funds, but the very legitimization of research flow, from the State.(9) And the bigger the obstacles confronting S&T progress, the more the governmental blessing and intervention are required.

Third, this conceptual and organizational model is perhaps not applicable everywhere because basic linkages are missing in some cases, but to many it represents both a normal and desirable situation, in which all current actions must be considered.

As anticipated, our thesis is that this all-embracing approach is both confusing and self-defeating. Relevant evidence for this will follow. But a point has to be emphasized now: we do not object to the legitimacy of specific policies for science and for technology in less developed countries (LDCs), Latin American included. On the contrary, we think that such policies are badly needed because no "invisible hand" will correct by itself secular insufficiencies. We just take exception to a certain style (i.e. holistic) of policy management prevalent in the area.

SCIENCE AND TECHNOLOGY: SOME MEANINGFUL DISTINCTIONS

The interplay between science and technology has never been smooth. Over centuries, science was highly dependent on other intellectual spheres (i.e. religion, philosophy), keeping incidental, and even contradictory, links with technology.(10) Among the ancient Greeks, for instance, technical dealings were a matter of disrepute fit for slaves and humble artisans. To some analysts this cleavage explains key faults of Greek science.(11) Furthermore, even on the verge of the Renaissance, those who tended to experiment with various work and methods (physicians, for example) frequently were assigned derogatory labels.(12) Although Bacon, and the tradition initiated by him, has established since then the social respectability of empiricism and technical concerns, convergence with science-based generalizations and practices had yet to be obtained.(13) This equivocal interplay continued until after the Industrial Revolution. Indeed, most technical innovations of this period were grounded more on a crude empiricism than on systematic research.(14) Since 1860, signals of a research-based technology appeared so that both elements give way to a cross-fertilization stage, which transformed science into a tool for explaining and changing the world.

Seemingly, this ultimate encounter between science and technology provided support to the holistic viewpoint. The industrialization of science, on the one hand, and the transnational expansion of research and its tangible outcomes, on the other, abolished any meaningful distinctions among different types and dynamics of knowledge inception and diffusion. Not so.

To begin with, the intimate articulation between scientific ideas and technical applications results from an aggregative survey which reflects the European evolution over the whole nineteenth century. But it is well known that countries (including the United States) differed in the degree to which they adopted scientific institutions and technical innovations.(15) The linkages were uneven in different countries and disciplines, though all (that is, Western Europe, England, United States, Japan, Russia) were affected by the "scientific revolution,"(16) and by a vigorous shift in production factors. True, one sees an ultimate coalescence of science and technology within the overall framework of the contemporary industrial system; however, distinctions between them remain meaningful on a national and sectoral level.

Second, the marriage between science and technology gave rise to new organizational patterns (i.e. the industrial laboratory subjected to governmental and entrepreneurial decisions. But this novel configuration took shape without excluding other forms of knowledge creation and dissemination. Academies and universities kept performing their classic functions; at the same time, there was room for individual inventions and technical improvements, divorced from systematic research.

Thirdly, from World War II on, there have been signs that this marriage faces painful tensions. A variety of factors, such as overspecialization, radicalization, and increasing research costs on the part of science; and the so-called barriers to entry, market constraints, as well as supply and transfer considerations on the part of technology account for them. This situation – whose content and components deserve a scrutiny which is beyond this chapter's objective – suggests that the linkage discussed so far is the result of singular historical and institutional circumstances. To be preserved and expanded, this linkage demands uncommon structural conditions that only a few countries can obtain.   For the rest (small industrial nations and LDCs) a sensate division of labor and an intelligent specialization seem the proper way.

Finally, science and technology are still sensitive to unlike factors. Thus, technological decisions and innovations are conspicuously affected by the size of the market, the type and level of industrial demand, the supply of entrepreneurial capability and skilled personnel, relative openness towards international trade, and the given economic policies. On the other hand, scientific activities are responsive to some components of the overall culture, academic arrangements, patterns and ways of collegiate recognition, and the differentiated dynamics of disciplines.(17)

This implies that scientists and technologists follow distinct patterns of recruitment, socialization, organization, and recognition. Only in exceptional cases may one apply the same yardstick to them.

A brief recapitulation is pertinent. The coupling of science with technology was neither easy nor universal; each has been responsive to, and has exerted influence on, diverse structural and cultural dimensions; and some links between them, though powerful and pervasive, do not represent an irreversible and expanding process on the national level.

If these statements are correct, the widespread approach to science and technology matters in Latin America should be revised.

SOME IMPLICATIONS OF THE ANALYSIS

Various hypotheses have been suggested in order to explain the relative scientific and technological backwardness of the area. One of them points to the Spanish culture as an inheritance uncongenial with the modern, scientific ethos; others refer to the type of economic evolution and organization (i.e. over-specialization, market size, excessive protectionism) as a deterrent to multifaceted and self-induced innovations; finally, the "neo-imperialist connection" is taken by many as a well-rounded explanatory category.(18) Most of these hypotheses need theoretical refinement and empirical substantiation. We shall not presently deal with them. Rather, attention will be given to the dialectics of this condition of backwardness with respect to the extensive holistic approach.

We have already argued that this dialectic undermines our understanding of the issues at stake. Let us examine four expressions of this effect.

To begin with, holism entails an undiscriminating view of the specific trajectory and needs of science and technology. As a result, there is a tendency to account for any scientific and technological gap according to a reductionist framework. In some cases, of course, reductionist devices are legitimate. But they are not a substitute for intrinsic explanations.   Premises and theories that have been put forward in order to interpret economic and social underdevelopment are also applied to the technological field. Thus, concepts like "structural heterogeneity," "internal colonialism," are mechanically and uncritically transferred to this field. As a result, important subjects – such as the evolving structures of the universities, their connections both with the power elites and the scientific centres, the cognitive and social standing of some disciplines, the patterns of publication and recognition, the intermediate role of scientific elites – are disregarded.

In other words, the holistic approach invites a reductionist explanatory framework which deters both the historiography and the sociology of science and of technology. It promotes only a partial understanding of the issues, subordinated to economic concepts and policies.(19)

Secondly, a too comprehensive view tends to oversimplify questions concerning the relevance of present-day science for developing countries.   Most authors think that these questions have to be settled by a blend of reasonable persuasion, fluid communication, and "right coercion."   But these voluntaristic precepts underrate the fact that the social estrangement of science in a typical Latin American context may be ultimately a self-protective mechanism, considering the tense relations between university centers and the state. On the other hand, more possession of know-how and techniques is not by itself evidence of scientific detachment towards development needs. Neither can it be remedied by lyric appeals or by spurious institution building.

Thirdly, the perspective now in vogue, i.e. blurring institutional boundaries among science, technology, and economics, overlooks the distinctive requirements of science as a fledgling product. Moreover, it disregards meaningful differences in recruitment, role formation, and long-range viewpoints specific to each sphere.

Finally, this view obscures problems related to the universalism of science. It is ordinarily argued that knowledge imported from abroad has had negative consequences because it is incompatible with the region's factor endowment and cultural traditions.   In consequence, science and technology should develop more "appropriate" contents, under the aegis of governments. This entails a peculiar jump from epistemology to politics.

In our opinion, the question of universalism carries a different meaning in science and in technology. It is important and irreplaceable in the former, while it admits variations and alternatives in the latter. Indeed, one needs to distinguish the universal contours of science as an intellectual adventure from the particular forms in which knowledge is embodied and transmitted. The search for a "special kind" of science undertaken by nationalist states spells an ideological exercise – and brings to mind painful antecedents – which is ultimately counterproductive.

In sum, the holistic approach contains four analytical defects. First, it hinders careful examination of the specifics of science and technology dynamics, limiting itself to reductionist (i.e. economist) explanatory statements. Second, it tends to put the blame on scientists for the scant utilization of modern knowledge and innovations in favor of the community, disregarding the responsibility of political elites and objective factors (i.e. overspecialization of the market, fragmentary intra- and intersectoral linkages) in this matter. Thirdly, it fails to encourage discrimination in the search for ways to foster scientific and technological talents, roles, and careers. And finally, it creates confusion in evaluating the universalism and particularism of scientific endeavors.

Certainly, these effects do not follow directly from the holistic approach. Different intermediate variables have a bearing on them. Our claim is that this approach encourages configurations along these lines.

Let us refer now to some of the self-defeating consequences of the holistic stand.

Moved by a conception that scarcely sets choice criteria, national councils for science and technology tend to put scientific and engineering groups in the same institutional and budgetary sphere. Under such conditions, these groups engage in fierce battles or convenient alliances which encourage the politicization rather than the "scientification" of decision making processes. In some cases, signs of social corruption surface through an overselling of services which these groups claim an ability to provide. Ultimately, neither the righteous nor the wise gain the upper hand, but, rather, the strongest and most cynical.

A second consequence of the holistic performance is more subtle. One of the features of science as a profession is an ability to bestow a sort of prestige, containing charismatic elements. In fact, the institutionalization and the productivity of science are inconceivable without them. We suggest that holism confers greater importance to political institutions and sanctions, at the expense of the scientific information role. No wonder that, under these conditions, men of initiative and talent tend to go after the source of power, either directly or after a short engagement with science.

Finally, holism pretends to impose an overall rationality which barely takes into account the uncertainties which surround scientific activities and technological decisions. With time it becomes a conceptual instrument too rigid to cope with new situtations.

CONCLUDING REMARKS

The design of science and technology policies in Latin America is a very recent process. For lack of tradition, experience, and industrial diversification, national bodies in charge of it have produced overambitious projects and expectations that have, perhaps unintentionally, negative consequences. It seems that selectivity has been sacrificed in favor of a self-flattering comprehensiveness.

Naturally, it is imperative to resist the undermining of scientific and technological growth in the area. So far history, politics, and the international environment have militated against it. But a grand conceptual design, dictated by an historical impatience with short-sighted political imperatives, is not the answer.

NOTES

(1) For a representative expression of this attitude see A. Araoz, "An Approach to Science Policy and Planning," in Organization of American States, Second Caribbean Seminar on Science and Technology Planning (Washington, DC, 1976), pp. 137-59.

(2) M. Herdia and A. Vega, "Toward a Science and Policy Organization in the Dominican Republic," p. 163.

(3) Ibid. p. 164.

(4) Republica de Nicaragua, Diagnostico Preliminar Sobre la Ciencia y lat Technologia en Nicaragua (Managua, March 1976), p. 6.

(5) Primer seminario sobre desarrollo cientifico y technologico de America Central, Organization of American States, (Washington, DC, 1976), p. 3.

(6) Consult P. Amaya and A. Alvarado, "La Variable Ciencia y Technologia en el Contexto del Desarrollo Economico, Social y Cultural," Ciencia, Tecnologia y Desarrollo, COLCIENCIAS, (Bogota, Colombia, 1977), p. 23.

(7) See, for example, the world and the regional Plans of Action produced by the Advisory Committee on the Application of Science and Technology to Development (ACAST), and also G.B. Gresford and B.H. Chatel, "Science and Technology in the United Nations," World Development, Vol. 2, No. 1 (January 1974).

(8) "Totalist" connotes a deep penetration of political institutions and concerns. It is not equivalent to "totalitarian," although in some cases certain parallels surface.

(9) See in this respect the Inaugural Address made by the Honorable Desmond Hoyte, in UNESCO, Consultation on Science and Technology Policies in the Caribbean Region, Georgetown, Guyana, December 12-16, 1977.

(10) For further details see Derek J. de S. Price, "Science and Technology: Distinctiona and Interrelationships," in B. Barnes, ed., Sociology of Science, (London: Penguin, 1972).

(11) On this subject see B. Farrington, Greek Science, (London: Penguin, 1944).

(12) For details see Sh. Pines, "Philosophy, Mathematics and the Concept of Space in the middle Ages," in Y. Elkana, ed., The Interaction between Science and Philosophy, (Atlantic Highlands, New Jersey: Humanities Press, 1974).

(13) See A. Thackray, "The Industrial Revolution and the Image of Science," in A. Thackray, ed., Science and Values, (Atlantic Highlands, New Jersey: Humanities Press, 1974).

(14) Evidences on this may be found in N. Roseberg, Technology and American Economic Growth, (New York: Harper, 1972).

(15) See J. Ben David, The Scientists Role in Society, (Englewood Cliffs, New Jersey, Prentice Hall, 1971), chapters 7 and 8.

(16) The term belongs to K. Boulding. See his paper "What went wrong, if anything, since Copernicus?" presented to the AAA's Symposium on Science, Development and Human Values, Mexico City, July 1973.

(17) See R. Merton, The Sociology of Science, (Chicago; Ill.: University of Chicago Press, 1973), esp. part 5.

(18) Some of them are reviewed by J. Hodara, "La conceptuacion del Atraso Cientificatecnico en America Latina: el Telon de Fondo," Comercio Exterior, (November 1976).

(19) Ibid.

14 Science and Technology Planning on a Regional Basis— The Central American Case

Jorge Arias
Francisco Aguirre

BACKGROUND INFORMATION

The five Central American republics (Guatemala, El Salvador, Honduras, Nicaragua, and Costa Rica) have an area of 441,000 square kilometers and a population of 17.8 million with a density of 40.4 persons per square kilometer.

The short-lived Central American Federation of States was created by the five Central American countries after they declared independence from Spain in 1821. Sixteen years later, for local political reasons, the five states disintegrated into five different republics. Afterwards, several attempts were made towards federations, the last one in 1921 when they were celebrating their first century of independence.

Nevertheless, by the end of World War II, the global trend for international cooperation was reflected in the interest of the less developed countries in strengthening their economic ties with one another in order to attain a better life for their populations. The Central American countries were no exception, and a movement was started toward the integration of their economies that would eventually lead to a united Central American economy. In 1951 a Program for Central American Economic Integration was prepared. Simultaneously, efforts in other fields were under consideration. Consequently the Organization of Central American States (ODECA) was created with the assistance of the United Nations, through the Economic Commission for Latin America (ECLA). Under the aegis of this movement, the initial Central American institutions were born as truly regional organizations. They were: 1) The Public Administration School (ESAPAC) located in Costa Rica, and 2) the Central American Research Institute for Industry (ICAITI) located in Guatemala. The latter was created in 1955 to conduct research for the utilization of local raw materials, develop manufacturing processes, diversify industrial production, and induce the utilization of modern industrial techniques. It was also empowered to conduct feasibility studies, to give technical assistance to existing industries and establish a Central American program on quality control. A few years later a Central American Bank for Economic Integration was created, and its headquarters were located in Honduras.

Some other regional institutions were also organized during this period, even though they were not part of the Economic Integration Program. They were the Central American University Council (CSUCA), formed by the five national universities, and the Central American Institute for Nutrition (INCAP), under the auspices of the World Health Organization.

In 1960 in Managua, Nicaragua, the Economic Integration Treaty was signed. It provided for customs unification and integration of industrial development, and also laid down the policy for transportation development. The signing of this treaty also marked the creation of the General Secretariat to administer the treaty and which, on a regional basis, is responsible for the correct application of the treaty and the implementation of the resolutions taken by the Economic Council and Executive Council, both created within the realm of this treaty. The General Secretariat (acronym in Spanish SIECA) is entrusted with responsibilities that place it on a quasi supranational status, since the other regional organizations (ESAPAC, ICAITI, ODECA, CSUCA, INCAP) operate previously established program actions in restricted areas of endeavor.

During the period of 1961-68, the economic integration movement showed very clearly the advantages that were to be gained in the process of development of the region. Naturally, differences arose between the participating countries, but the advantages and the increasing interdependence became obvious. The industrialization process gained great impetus and the financing mechanisms were able to operate on a more ample basis. The intraregional trade grew dramatically, and it became evident that the trends in development that were being achieved were of a higher rate than if the five countries had each gone their own way independently of one another.

One of the problems that became evident during this period was that, though the region as a whole was developing satisfactorily, the same was not true for each one of the partners. Some of them were advancing more rapidly than others. To complicate matters, at the end of the decade, an unfortunate border war took place between two of the countries.

The implementation of the integration program went faster than could have been envisaged when it was designed. Therefore, the need arose to start thinking about a more ample common market that could cover not only trade, infrastructure, and industrial development, but also other matters related to social and agricultural development. It was also pointed out that there was a need to accelerate the decision making procedures and to find formulas that would distribute, on a more equitable basis, the benefits, as well as the costs, of the integration.

The General Secretariat (SIECA), with the assistance of the Latin American Institute for Integration (INTAL), prepared a program for an ambitious Central American integration during the next decade. With this background, the Governments established an ad hoc, high-level commission for the improvement and restructuring of the Central American Common Market. This commission, after exhaustive consultations and discussions, produced a proposal for the creation of a Central American Community that could represent a valid step towards an ultimate political reunification of the five countries. This proposal, very thoroughly presented, contemplated widening the scope of the integration movement and institutionalizing several regional mechanisms for the bureaucratic operation of the process. At present, this proposal is being reviewed by each one of the countries. The natural political question is when and how the next step should be taken.

In spite of the above mentioned problems, the Central American Common Market continues to operate and develop, as a result of both the inertia and the impetus of its initial decade of existence. Just to mention a few indicative figures, in Central America as a region:

- Foreign trade (export and import $6.5 billion) ranks third in Latin America (after Brazil and Mexico). It is higher than Argentina and more than double the foreign trade of Colombia, Peru, or Chile.

- The exports of the Central American region represent 40 percent of what Austria or Spain export and are larger than those of the Dominican Republic, Bolivia, Ecuador, Paraguay, and Uruguay put together.

- Central America is already the second greatest exporter of coffee in the world.

- Meat (hoof-and-mouth disease free) export from Central America ranks number one in Latin America.

- The world consumption of bananas is supplied by Central American production, as the number one region of the world for export of this commodity.

- Traditional exports, that 10 years ago represented 85 percent of the total, today have lost importance relative to the nontraditional exports, which now represent 32 percent.

- The intraregional trade grew from $30 million in 1960, to $384 million in 1973, to $609 million in 1976, and passed the mark of $700 million in 1977.(1)

- The Gross National Product went from $2.6 billion in 1960 to $11.5 billion in 1976, with a GNP per capita growing from $242 to $650 in the same period.

These figures are not only indicative of the energetic development that is taking place in the area, but are also a measure of the challenge of improving, in real terms, the life and well-being of the population. The general feeling, widely scattered in Central America, is that the five can do better together than individually, but that equitable mechanisms have to be implemented in order for the benefits to reach all Central Americans.

SCIENCE AND TECHNOLOGY IN CENTRAL AMERICA

Scientific and technological activities in Central America have not had a high level of development as evidenced by the 1970 ICAITI survey.(2) This survey showed that, during that year, the total investment for research and development, as well as for dissemination and other activities conducted by research institutions, totaled only $25.4 million. Within this figure, a total amount of $9.3 million was devoted to R&D, while the major part of this sum was dedicated to applied research ($8.6 million). Only $300 thousand was being applied to pure research and $400 thousand to development activities.

If the above figures are related to the GNP and per capita, there clearly emerges a picture that is very bleak in comparison to developed countries but similar to other LDCs. Numerically speaking, this picture is as follows for 1970:

	Expenditure in R&D	
Country	% of GNP	$ per capita
El Salvador	0.27	0.81
Costa Rica	0.15	0.81
Guatemala	0.17	0.57
Honduras	0.23	0.57
Nicaragua	0.07	0.29

If these figures are compared to similar ones from countries like Argentina and Spain, one can see a similarity, because these two countries spent on R&D, in relation to the GNP, 0.28 and 0.2 percent respectively. One could infer that what is important is not the relative significance, but the minimum level, of the total expenditure. The per capita R&D expenditures of these two countries were $2.1 for Argentina and $1.9 for Spain.

If the total amount of expenditure on R&D in Central America is allocated by sector, the low level of expenditure by the private sector becomes obvious. The majority of the expenditure takes place in the public sector, with lesser amounts in the education sector, even though there are variations from one country to another. Government expenditures represent 50.5 percent of the total, whereas regional institutions account for 25.5 percent and national education centers, 15.6 percent. Since these sectors are directly or indirectly government related and add up to 91.6 percent, it becomes clear that there are possibilities for a better coordination of science and technology activities through a well thought out policy.

As expected, the major portion of the R&D effort is dedicated to agriculture (36 percent). Medical sciences represent 19.6 percent, mathematics, natural science, and engineering, 17.6 percent, and social sciences (principally economics) 12.6 percent.

The private sector, as stated above, has had a smaller relative importance in these activities. It normally lacks research facilities and usually contracts for its needs with research institutions that, as indicated before, are government-related.

The weakness of the science and technology sector in Central America is a major factor in the lack of penetration of modern technology into the economies of these countries. But it is not the only factor. It is important to realize that if the availability of S&T is weak, the demand also tends to be small. This is due, in part, to the fact that many industries are subsidiaries of foreign companies that do not depend upon local technology to solve their problems. They have their own resources for this purpose.

In general terms, one could say that the main factors that appear as stumbling blocks, preventing local science and technology from becoming a major input in the process of socioeconomic development, are associated with the underdevelopment syndrome and, therefore, the vicious circle becomes more evident. Some of these factors are the following:

1) Low rate of capital accumulation due to the low levels of savings and investment. It has been estimated that the fixed capital available on a per capita basis barely reaches $1500 with an annual rate of increase of only 2 percent. This last figure is only a third of the average for Latin America. For the year 1973, it was estimated that the increase in fixed capital was only $74 per capita, which is half the amount shown by the other Latin American countries.

If the rate of accumulation of capital is limited, so is the acquisition of new equipment and machinery, which is one of the ways of transferring modern technology into the productive system of a region.

2) Industrial underdevelopment. The Central American countries have traditionally been devoted to agriculture, and only since the common market was implemented, 15 years ago, have they started intensifying their industrial development. Nevertheless, the industrial output in its contribution to the GNP is still low – 17 percent for 1973, and absorbing only 10.9 percent of the labor force. By contrast, the contribution of agriculture that year was 25.8 percent and absorbed 51.6 percent of the labor force.

The industrial sector is predominantly of the traditional type (food, beverages, textiles, and tobacco), where the added value was $876.7 million (75 percent of the gross industrial products) in 1973. More modern industries (pulp and paper, rubber, petrochemicals, and nonmetallic minerals, had an added value of $290.4 million during the same year. Nevertheless, the more modern groups have grown fast. In 1960 they had contributed to the gross industrial product only $45.9 million, 12.3 percent of the total. In 1973 they contributed 24.9 percent of the total.

Modern industries are normally the main consumers of R&D and also influence very much the spread of modern technology. Therefore, the continuing underdevelopment of this sector is a major reason for the lack of integration between the industrial sector and the infrastructure of science and technology.

Despite the initial growth of the modern industrial sector, local demand for the output of modern technology is still limited, and is covered mainly through imported goods. Imported goods command high technology input, and export products are mainly raw materials of low technology input, which is another example of the technological gap that has a negative effect on the demand of R&D from the local S&T apparatus. This fact, added to the already-mentioned factor of close relations between subsidiaries or branches of foreign firms which make a very low demand on local R&D, together with the introduction of complete turnkey jobs of "nil" requirements for the same, account also for the low level of integration between industry and R&D facilities.

3) The size of local markets, the uneven distribution of income, and the low educational level are other factors that are affecting the development of science and technology.

The size of the market is still small by comparison to the developed world. It is true that the population is growing fast; as a matter of fact, during the last 15 years, it grew at a rate of 3.1 percent bringing the total number from 11.8 million in 1960 to 17.6 million in 1975. But this fast population growth led to a decline in per capita income, and has helped to accentuate uneven distribution of same. Therefore, a great part of the population is only barely above the subsistence level, with very little change in consumption of high technology products. The educational level is also a negative factor. For the population of 10 years of age and above, the censuses of 1973 have shown a level of 43 percent that do not read or write (with striking variations between countries such as Costa Rica with only 11 percent and Guatemala 52 percent) and a low rate of school attendance.

At the university level there are six national universities and seven private ones with a total enrollment of 76,000 students in 1972 (12 percent in the private universities). The university enrollment has grown during the last eight years at a rate of 16 percent per year, and the largest group (30 percent) goes into law, economics, and business administration, whereas engineering-related fields attract only about 14 percent. During 1973 the graduation roster of 2,400 professionals showed the following picture:

FIELD	NUMBER	PERCENT
Engineering and Architecture	366	15.1
Medicine	310	12.8
Law	263	10.9
Economics	252	10.4
Dentistry	94	3.9
Agronomy and Veterinary	56	2.3
Chemistry	40	2.0

In this table such fields as psychology, education, and philosophy are not included since the available information did not permit a reliable classification. It points out, anyway, that for the early 1970s fields related to science and technology had a low level of participation in relation to the rest. The traditional system of Latin American university autonomy is also reflected in the lack of integration among the university system, the governmental R&D establishment and the productive system. Unfortunately, this gap also affects socioeconomic planning as it relates to science and technology planning, since the former, in a way, dictates the priorities of the latter.

During the past integration process in Central America, science and technology have played a secondary role; economics has been the main subject, and presently is being expanded to include social aspects. The trend is starting to reverse now. Costa Rica has created the National Council for Scientific and Technological Research (CONICIT), and Guatemala included in its National Plan for Development (1975/79) a chapter dedicated to science and technology that calls for the creation of a National Committee for that purpose. The National Planning Offices have created science and technology units.

A major impact on the scientific and technological development of the area could have been affected by the Central American University Council (CSUCA), organized in 1961 to integrate the professional education of the area. Initially, the master plan for basic sciences offered a great opportunity for the improvement and enhancement of a sound program in the principles that underlie the fields of science and technology. Moreover, this plan opened the road for postgraduate training at a regional level in order to make better use of the human and financial resources available. Indirectly, this would have also helped in the professional integration of the area.

The achievements in this plan have been, unfortunately, short of the expectations. In terms of research, under this plan only a regional study of land tenure and labor agricultural conditions was conducted through the Central American Research Institute of Social and Economic Science. Recently, a second project has been implemented for the study of internal migrations. In sanitary engineering, a postgraduate regional school was organized in Guatemala which has been successful in the teaching aspects but, because of lack of funds, has not implemented its research program. These examples show that CSUCA, in spite of its potential for science and technology development, has limited its scope of activities so far to the training of human resources.

On the other hand, the scientific regional institutions have proven more effective in the implementation of their original plans. Two of them, the Institute of Nutrition of Central America and Panama (INCAP) and the Central American Research Institute for Industry (ICAITI), described above, have been very successful not only in conducting research for the area, in their respective fields, but also in creating the proper environment for the training of professionals in research techniques, and in disseminating the results of their work through publications, seminars (short of formal courses), at the undergraduate and postgraduate level. The main problem that these institutions face is that when they attempt to carry out some of their regional programs, they find that the countries have not always decided their local policy at the national level and thus force the institutions to decide by themselves on the best course of action.

Mention should also be given, in this chapter, to the Interamerican Institute of Agricultural Sciences (IICA), located in Costa Rica. Even though it operates on a Latin American basis, its presence in Central America has been of decisive influence on the agricultural development of this area. It is at present being transformed into a Tropical Agronomic Center for Research and Education (CATIE) associated with the University of Costa Rica.

Some other organisms have been created under the realm of the integration scheme that are only collaterally related to science and technology. One of them, for example, was the Permanent Commission for Agricultural Research created in 1965 by a joint meeting of the five Central American Ministers of Economy and Ministers of Agriculture. This Commission was entrusted with the responsibility of designing a policy for agricultural research, at both the national and regional levels, and to promote the integration of the research activities going on in the area. It was also in charge of preparing a master plan for agricultural research in the area. Unfortunately, the accomplishments of this Commission have not lived up to expectations..

The master plan for an integrated development of the area during the present decade, prepared by SIECA/INTAL, does not have a special reference to specific science and technology policy, even though some of its considerations could be considered as making reference to such a policy. For example, when it makes reference to foreign investment related to industry, it indicates that the lack of accurate information prevents the determination of a sound policy that could promote the optimization of the contribution of foreign investment towards the industrial and technological development of the area. This plan also makes reference to the challenge of selecting adequate technology for the region, to the implementation of proper technological research programs, and to the evaluation of the job-generating capacity of the available technologies. This last particular aspect is of great importance to the area in view of the fact that the rate of growth of the population is still very high.

This master plan also calls attention to the fact that it is very important to establish an educational program coordinated with research activities and industrial development processes, as well as with agricultural and service development. In this respect it clearly states "a regional policy for technology development should be established that should aim at increasing the engineering capacity, not only as an academic discipline, but more so as an instrument for the integrated development of the area."(3) The plan recognizes the need to utilize foreign technology, but also emphasizes the importance of acquiring the domestic capacity necessary to evaluate the available technological alternatives, in order to facilitate wiser decisions in determining the most economical and efficient choices for the particular circumstances of the area, taking into consideration the balance-of-payments, patent rights, know-how, and licensing normally associated with these technologies.

PERSPECTIVES OF A REGIONAL
SCIENCE AND TECHNOLOGY PROGRAM

The background information and descriptions given in the previous pages could be regarded as indicative that the time is ripe to design and implement a regional policy for science and technology. The different officials who have met to plan the restructuring of the common market have repeatedly expressed their interest in incorporating S&T as one of the instruments for the future development of the region and, furthermore, have expressed their concern over the problems related to the process of transferring technology, especially in regard to the costs involved, the negative effects on the local development of the S&T system, and the lack of coordination of this system with the public and private sectors. This preoccupation is valid not only on a regional basis but also at the national level in each one of the countries. Costa Rica has already established its National Council For Research (CONICIT). Nicaragua has prepared a project for establishing a similar organization; and in El Salvador, at the National Planning Office, they are also contemplating a similar one. In Guatemala, the Development Plan 1975-1979 specifies that the Secretary General of the National Planning Commission should promote meetings with the other national organizations of the area to discuss a unified science and technology development policy for the region. Further, it states the importance of establishing priorities and defining concrete goals, as well as means for achieving them. Special emphasis is given to the establishment of a well-coordinated information service.

All these ideas have been gathered and incorporated into the proposed treaty for the establishment of the Central American Community presently under discussion. This project contains elements of policy and programming in areas such as agriculture, industry, food, energy, transportation, education, health, housing, and labor that reflect the intention to develop science and technology. In spite of this, there is a special chapter on the commercialization of technology, and the application of science and technology to basic planning for the development of local science and technology. In accordance with this project, the Council, which constitutes the highest organ of the Community, should adopt a common policy for the commercialization of the technology and its application for development, in accordance with the best interest of the region, and guarding against an undue encroachment on the autonomy of the region in these matters.

Some of the policy elements established in this projected treaty recommend, a system to monitor, on a continuing basis, the acquisition of foreign technology, particularly its cost and limitations; the strengthening of the domestic capacity for science and technology; the promotion of basic and applied research, especially in relation to the production capacity of the area; and increased use of science and technology policies for the economic and social planning of the region. Finally, the system hopes to promote interaction between the Community and other countries or regions of the world, in order to attain a more efficient use of science and technology in order to improve the quality of life in the region, to lower the level of unemployment, and to improve its productive capacity through a better use of local resources, with due respect for preservation of the natural environment.

The above statements are a clear indication of the recognition that, in the process of integration of Central America, science and technology are to play an increasingly important role, and that the assessment of S&T for its socioeconomic impact, will become an important issue in agricultural and industrial development of the region.

The projected treaty has some general guidelines for the commercialization of technology. Registration and authorization of contractual arrangements for transfer of technology are contemplated; special care is given to the avoidance of restrictive clauses that curtail the initiative of the user or the objectives of the treaty ; and a system for industrial property and trade marks is outlined. Finally, the projected treaty envisages the establishment of a Central American Council for the Application of Science and Technology (CONACIT), in charge of formulating and implementing the science and technology policy of Central America. It also makes reference to the establishment and/or strengthening of local institutions charged with responsibility for conducting programs in applied research, dissemination, and development of science and technology. Standardization, technical services, and information facilities are also contemplated within the realm of the S&T policy.

The operational part of these institutions is analyzed in terms of financial resources, through the establishment of ad hoc funding mechanisms or foundations that can give proper economic backing. All these efforts should also take full advantage of international cooperation. Necessary legal actions to implement the above are also envisaged to assure proper compliance with the policies that will be designed.

Research institutions already in existence such as ICAITI (industrial research), INCAP (nutrition), ICAP (public administration), CSUCA (university affairs) – and those to be established – such as ICAITA (agricultural research), INCOME (foreign trade), INCEMSO (occupational risks and labor health) and CONACIT (science and technology council) – will be the institutions of the Community, regarded as institutes or specialized agencies.

MAIN ACHIEVEMENTS

ICAITI, in anticipation of what is to come, has already prepared some basic studies to help in the preparation of the guidelines. Some of these refer to the gathering of basic information, such as the OAS-assisted survey of Central American resources that can be utilized for science and technology. ICAITI also participated as a subregional focal point for the Pilot Project for Transfer of Technology of the OAS, and has continued its specific tasks under this project, on iron foundry and related aspects. Under the auspices of INTAL, there was also a survey of the industrial characteristics of firms that utilize foreign technologies. For this specific project, the Federation of Chambers of Industry (FECAICA) has provided valuable cooperation.

With the assistance of the International Development Research Center (IDRC) of Canada, ICAITI also conducted a seminar on the science and technology development in Central America. This seminar was conducted in two parts: the first with the participation of international and regional institutions, and the second with the local national institutions (ministries of economy, planning offices, science councils). When this seminar was conducted, the Community treaty draft had not been finished, and specific guidelines were not available. Nevertheless, the basic elements for an S&T policy had been outlined, and the government and private sectors had already expressed, one way or another, recognition of, and interest in, the input of science and technology into the process of socioeconomic development.

The course of the seminar demonstrated the importance of analyzing the supply and demand for technology, not as a mere passive element of the implicit policy of development plans, but rather as an integral part of the socioeconomic process of transformation. Therefore, the importance of specific guidelines for S&T development in the overall policy was recognized. Participants also recommended that care should be taken to acknowledge any changes that might take place locally or at the international level. The importance of human resources, trained for the purpose, was also recognized at this seminar.

Nevertheless, it was acknowledged that the simple expression of general objectives constituted little progress towards attaining goals, that it was necessary to spell out specific science and technology planning action, making use of local and international resources. Two main avenues of progress were identified: strengthening the already existing science and technology system to promote expansion of demand, relating it to the local supply of technology; and identifying and implementing Central American projects for technological development which were related to the concrete technological demands of those activities deriving from the integrated development strategy of the area.

In this regard, it was recognized as important that the institutions should conduct basic studies of several subjects, including:

1. the science and technology system of each one of the countries, especially its resources, capacities, and potentialities of expansion;

2. requirements of science and technology inputs into the integrated plans for development;

3. supply and demand of science and technology human resources;

4. mechanisms that Central America could utilize for the promotion of science and technology;

5. characteristics of the present process of technology transfer;

6. public sector purchase and acquisition policy, and assessment of its use as a mechanism for the promotion of industrial, scientific, and technological development;

7. those circumstances that constrain full utilization of the results of scientific research, and promotion of correlated activities between research and development;

8. design of an information system basically related to science and technology development, which would include both industrial services and, if feasible, a data bank.

Additionally, the seminar recommended the identification of a series of projects of a technological development nature, that would become evident from the priorities of the integrated development plan of the region. It was recommended that existing institutions be utilized to carry out these studies with participation by international experts when needed. Specific mention was made of the food, paper, and pulp industries. Emphasis was placed on conducting a project that would entail the "disaggregating" of the technological package.

Finally, the seminar recognized the need for preparing technical people in the area of planning and policy design of science and technology, and also recommended that technical coordination meetings should take place for proper exchange of ideas among those people in charge of implementing the projects.

There is not, as yet, a final program for implementing all the recommendations of the seminar, but some of the actions that have been started already are:

- Work is already underway, under the auspices of the OAS Program of Science and Technology for Development, for the transferring of technology in the development of the metal-mechanical industry in Central America.

- Discussions were held with the Inter-American Institute of Statistics on the methodology of preparing science and technology statistics. The methodology has since been tested.

- In the planning stage are several studies to evaluate the purchasing policies of the governments, as well as consultant and engineering services available in the area, and the current mechanisms used for technology transfer.

- A course was organized and held in Costa Rica, on a regional basis, on science and technology policy   design, and some other similar courses and seminars, especially at the national level, have already taken place.

- Various meetings for coordination of the information services have already been held, and the creation of a Central American network of information and technical assistance services is underway.

All basic studies, as well as the Central American projects for technological development, are being planned to include both local personnel and foreign experts. In this way, a training element will become a very valuable by-product of the effort. Of further importance is the fact that, in July 1975, when the Ministers of Economy met and were informed of the results of this seminar, they approved a resolution backing the recommendations presented.

Finally, it is worth mentioning that even though Panama is not now a member of the integration scheme, it has expressed interest in becoming one because of its recognition that its problems are very similar to those of the other Central American countries.

RECENT DEVELOPMENTS

The Interamerican Committee on Science and Technology (CICYT) in its fifteenth meeting (November 1974) took note of, and endorsed, the document entitled "Outline of a Central American Program for Science and Technology Development." This document described the results of the first Central American Seminar on Science and Technology Development. The main components of a Central American program for this purpose were identified and a tentative budget for implementation was also included. When this program was endorsed by CICYT, and later by the Interamerican Commission on Education, Science and Culture (CEPCIEC) the initial steps were taken to obtain at least partial financing from the International Development Research Corporation (IDRC) of Canada. Unfortunately, the result was negative.

The Interamerican Council of Education, Science and Culture (CIEC) held a meeting in 1975 in Mexico and adopted resolution CIECC-242/75 which, taking into consideration the advances already achieved, decided to support the Central American countries in their efforts to prepare a science and technology plan that would contemplate: 1) basic orientations for expansion of technology supply and its correlation with demand; 2) identification of Central American projects for technology development that should correspond to concrete high priority needs of the industrial sector, and 3) optimization of the use of Central American institutions for local and regional needs in order to attain the foregoing goals. Furthermore, it was requested that the General Secretariat of OAS, as well as other more developed countries, should also provide assistance. Finally, it was recommended that the Central American countries should set up an Advisory Committee to supervise the implementation of the plan and to submit an action program on the basis of the results.

After some preparatory meetings, the Advisory Committee was established and held its first meeting in San Salvador in January 1976, with two representatives from each of the countries involved. This meeting established the General Rules and Regulations of the Central American Commission for the Development of Science and Technology. A main goal was established for this Commission which stated "that on the basis of priority programs of the countries, and with the guidance of the regional institutions the common basis for a development plan should be identified so that science and technology would actually contribute to the social and economic development of the Central American Region."(4)

In this meeting, it was also decided to request the governments to: 1) define a national science and technology policy and design their plans for science and technology development;   2) strengthen the institutional structure for planning, coordinating, and implementing science and technology activities; and   3) analyze the different   plans and activities in science and technology going on at present in the region, so as to assure a close relationship with the plans and priorities of their respective countries.

As a result of the interest expressed in this meeting, the Central American countries have created Units of Science and Technology within their Planning Offices, with the exception of Costa Rica which already had a National Council for Scientific and Technological Research (CONICIT). At present, the Central American Commission of Science and Technology Development is mainly   constituted by representatives of these units. After the meeting in San Salvador, the Commission has held three additional meetings: Costa Rica (1976), Guatemala (1977), and Nicaragua (1977).

During the meeting in Costa Rica, there was discussion of several strategies for coordinated action on a regional basis. But a closer relationship between the Commission and the existing research centers, which would make implementation of the selected projects easier, is still lacking. During this meeting, a proposal was discussed to develop the necessary mechanisms for working up a planning policy for the region. After further refinement, the proposal was approved, during the meeting in Guatemala, and initial steps have been taken to secure financing from the IDRC (Canada) and the Central American countries for its implementation.

This Commission also has been discussing the participation of Central America in the United Nations World Conference on Science and Technology, and during the meeting in Guatemala the initial steps were taken to prepare the national documents as well as a regional one.

It is also importat to point out that several additional actions have been taken, including the Technical Information Service (SIATE) currently being implemented under the auspices of the OAS Mar del Plata program, which also includes the Dominican Republic. Under the coordination of ICAITI, the five Central American information centers wil provide more efficient service to their Central American users. Several seminars have been held for this purpose. A by-product of this activity is the plan to publish, collectively , a catalogue of periodicals and directories of commercial representatives, official translators, etc. Work is also underway to prepare norms and standards for documentation services, exchange mechanisms for question-and-answer services, and other related activities that would tend to coordinate the limited resources available into a more efficient overall system.

CONCLUSION

There is no firm policy for the scientific and technological development of Central America. Nevertheless, background information is being accumulated along with supporting data and evidence. The lack of concrete national policies for all countries both hinders and facilitates the development of a regional one. If the technological policy has as an aim objectives for the socioeconomic development of the area, it certainly would coincide with the national policies and, thereby, will make easier the design of such a policy on a regional level. There is hope that the development of the project on Instruments for Scientific and Technological Policy and Planification could be the beginning of a new and successful effort.

There is much need for such a policy because of the weakness of the science and technology system of the area. On a cooperative (regional) basis, it is likely that a more efficient pool of resources can be put together and, thereby, attain, in a shorter period of time, the objectives being sought. Autonomy in the decision making process for technological matters is highly desirable in order to have the capacity to identify the true problems, seek alternative solutions, and select and implement the more practical ones for the area. To undertake this task implies recognition that regional actions cannot be totally divorced from national policies but, on the contrary, must be highly coordinated.

Any technological policy for the area should be complemented by a wide planning for scientific development that includes research programs of pure and applied nature, leaving the former, together with education, for the university system.

Science and technology will continue to play a very important role in the process of integration of the five Central American countries. There are certain unavoidable activities that require regional size for implementation, either because economies of scale are essential, or because they attack problems shared in common (ocean resources exploitation, tropical woods, environmental contamination, use of hydrological resources).

- It is important to coordinate the national policies of industrial property, foreign investment, and standardization to make them effective in the process of integration.

- It is to be expected that the experience gained during the more than 15 years of existence of the common market will be wisely utilized, so that the restructuring currently taking place will truly result in a Community with higher quality and quantity of life for all of its inhabitants. There is, at present, a feeling that any process of economic integration should be accompanied by technological integration as well, so that the benefits of a more ample market are not diminished or even worse, lost because of lack of technical capacity to take advantage of the same.

NOTES

(1) Preliminary data for 1977 as announced by SIECA in El Imparcial, January 9, 1978, Page 5.

(2) ICAITI "Survey of Scientific and Technological Activities in Central America," Instituto Centroamericano de Investigacion y Technologia Industrial, Guatamala 1970.

(3) See for example Tratado General de Integracion Economica Centroamericana – Revista de la Integracion Centroamericana Vol. 1 Junio de 1971.

(4) See Comision para el Desarrollo Cientifico y Technologico en Centroamerica y Panama, Informes de las reuniones celebradas en Salvador, 1976.

15 Science and Technology Planning Problems in Small Caribbean Countries
Arnoldo K. Ventura

The standard of living disparity between the rich and poor societies continues to widen at an ever increasing rate.(1) All social and economic maneuvers over the last two decades to narrow this gap have failed to create more equity between developed and developing nations but, instead, have aggravated the situation between rich and poor nations, as well as deepened the chasm between rich and poor within national borders.(2) This is reflected in the fact that although the per capita GNP of some countries rose significantly, there was some improvement in science and technology (S&T) infrastructures, neverthe- less, a third or more of the populations of LDCs still languish in poverty and dispair with inadequate nourishment, shelter, clothing, health care, occupation, and without any visible means of existence.(3) The type of democracy which prevails in many of these countries seems to foster a distributive system which anchors the majority in illiteracy, sickness, and social indifference, thereby alienating them from the political process, while a minority revel in luxury and remain totally oblivious to the human deprivation and despair around them.

In the face of these inequalities, an abundance of technological energies and skills are applied to the production of consumer goods of marginal importance, armanents and other paraphernalia of national prestige. Some 40 percent of the world's expenditure goes to national prestige and defense, but less than five percent is directed at the problems of two-thirds of mankind who strive to survive on a meager agricultural base.(4)

These realities brought into sharp focus the need to marshal S&T to serve the majority of mankind, and the United Nations and other international bodies have taken steps to promote the political will and structures necessary for the redeployment of S&T to the pressing humanitarian issues.(5)

There was need for legislative and executive measures in most developing countries to improve the resources for S&T, and to provide the innovations and skills to meet national goals for productivity. It is now generally accepted that a clearly enunciated science policy is vitally necessary to strengthen research and development (R&D) capabilities and to link them with the productive political and educational systems.

Before the problems experienced by small Caribbean countries in formulating S&T policies can be effectively discussed, it appears necessary to indicate the characteristics of such policies, and to elaborate further why they are being contemplated in many of these countries.

Before World War II, the need to manage science was not fully appreciated, and indeed was opposed by the socioeconomic systems prevailing to western countries.(6) The need to organize S&T towards social objectives became more obvious during the last 40 years, which witnessed spectacular successes of S&T in attaining economic and political power.(7) Once the idea of technology as the major driving force of socioeconomic progress was appreciated,(8) countries began investing an increasingly large portion of their physical, financial, and intellectual resources in these activities. Further, the effectiveness of scientific knowledge in all aspects of public life has been recognized as being so pervasive and powerful that it was considered politically inexpedient to allow scientists alone to determine the consequences of their work.(9) All those countries which recognize S&T as an indispensable precondition for national progress, took steps to formulate codes of conduct and planning for S&T. These codes took the form of national frameworks for legislative and executive action to link S&T activities with education, employment, and production to achieve specified social goals, with the hope that these actions would eventually create scientific traditions for these countries. These maneuvers constitute S&T policies.

The incisive consequences of S&T activities on socioeconomic and cultural development, and the need for S&T policies became clear to most developing countries only within the last decade. At this juncture, most states came face to face for the first time with the intricacies and difficulties encountered in making science policies, and realized their weaknesses and lack of confidence in making decisions with regard to such policies and their impotence in implementing them.

The existence of severe problems of hunger, transportation, health, housing, education, and clothing in many countries, while S&T remained uncoordinated and fragmentary, were clearly recognized as a travesty of justice and an insult to humanity by the international community. However, it was also realized that these ailments are vast, complex, and fraught with political overtones, and could only rationally be handled by planned, collaborative, concerted actions focused on well-defined objectives. As the body of scientific knowledge grows, and the complexity of the social problems become better understood, it becomes obvious that multidisciplinary action is necessary. The task of welding the S&T specialists and institutions into a cooperating unit to identify, formulate, and execute specific projects to tackle the varied socioeconomic problems, and to solicit national and international support for these solutions are the great challenges of science policies.

CHARACTERISTICS OF SCIENCE POLICIES

Having examined the basic concepts and utility of a science policy, it would appear instructive to outline in greater detail certain universally accepted objectives of such policies to allow better recognition and identification of root causes for some of the problems encountered in formulating and implementing these plans.

Until relatively recently, it was felt that if science policies were necessary, they should be restricted to development of science per se and should not be concerned with the utility of science.(10) In many developed countries, even the notion of the necessity to provide for science was not well accepted, because it was perceived that science, because of its intrinsic qualities, would always be respected, and, as such, would always be supported. By mimicry a similar stance was also adopted by many developing countries. This has proven to be a sad mistake in both the developed and developing countries, especially in the latter, where scientific institutions were supported primarily because they could enlarge national prestige in a somewhat cosmetic fashion, and consequently was the first to be sacrificed in stringent economic periods.

It is now clear that science policies must seek to promote science per se, as well as provide the conditions for the coupling of research and development with the productive process. In small countries, there is no question that science must be managed in such a way that the limited resources are maximally employed in production. This is not to say that basic research should be neglected, but it should take place with a very watchful eye on the careful development of local natural resources. To achieve these ends, well enunciated policies must be formulated to provide a rational basis for action.

The basic aim of S&T policy is to channel all the relevant scientific activities towards national development, i.e. toward growth and change. These aims will have to be translated into specific S&T objectives. To achieve growth and change for socioeconomic progress, the objectives must then be prioritized and categorized into those to gear the scientific community and the lay public to innovate, and those to stimulate production. Once the necessary conditions for research and development are provided, and incentives for creativity identified, there is not much more that can be done to stimulate innovation since this commodity is not directly linked to any form of natural resources except, probably, brainpower and attitude.

The second aspect of a policy is the application of S&T to production. The modernization of the productive infrastructure for more efficiency and social acceptability by the introduction of modern, local, and foreign techniques into priority areas (such as agricultural and industrial development with special emphasis on small businesses, rural and urban upgrading, agrarian reform, trade, and employment policies) is vital.

Science and technology have profound effects on the dual aspects of development – growth and change. They can boost productivity and contribute to a new social ambience. A main feature of science policy is to stimulate change in social, economic, cultural, and educational sectors, in other words, cultivate a national scientific tradition. For many developing countries, the single most important contribution of science may be the introduction into the national psyche of scientific attitudes such as clarity of thought, honesty of purpose, accuracy in assessments, and an allegiance to excellence. In this respect, the interphase between the scientific community and the public at large must be made vibrant and reciprocating. Science planning must first seek to quantify the resources available for its activities. For each countries, critical mass of brains, equipment, facilities, finances, and specialized institutions are necessary if there is to be a flow of technology into production. A science policy, irrespective of how well conceived, will not be satisfactorily promulgated nor implemented if the adequate infrastructure is not present to instrument and rationalize its goals. There must be a centralized, coordinating body acting as the pivot of the infrastructure capable of digesting the statistical data, of advising and monitoring, while encouraging the use of policy instruments. Reconciliation of the competing claims for resources and national attention must be undertaken by a well-defined and respected scientific entity.

The development of indigenous science and technology capabilities, as a matter of great urgency, must be tackled by S&T policies in developed countries. Without this core of expertise, the ability to exploit local natural resources and to enable proper transfer of technology will be inefficient at best, if not impossible.

Science policy must contribute also to the national will by way of recognition and sensitivity for S&T so that these imperatives may find a way into production in the shortest possible time. Science and technology must be popularized in all the nations, and incentive provided for its use. An appreciation of science will eventually mean recognition of the scientist and his work.

A most difficult problem facing all science policies is the economic analysis of science. Technological merit, scientific merit, and social merit(11) are all parameters which have to be considered in evaluating science. Although it is difficult to quantify any of these variables, policies must somehow provide some guidelines for their assessments.

All S&T policies eventually will have to place in clear context the value of maintaining ecological balance against economic criteria. This is extremely important for small, island countries.

Because science is universal, the S&T plans of each country should include directions towards international cooperation and sensitivity.

REQUIREMENTS FOR IMPLEMENTATION OF SCIENCE POLICIES

Having provided the ingredients and ultimate aims of S&T policies, it would then be useful to examine the basic requirements for the support and implementation of such policies if they are to be effective.

For any such policy to be effective, there must be a keen awareness of the usefulness of science and technology within the community at large. The demand for S&T should be sufficiently expressed by the community, and its relevance clearly understood. Also, for any policy to be effective, it must have the full backing of the custodians of the society in which it will be implemented.

Since these policies concern the use of new methods and procedures, the users must be sufficiently acute and amenable to innovation, and be confident that the necessary technological services (such as standards monitoring, marketing, extension services in agriculture and industry, as well as specialized personnel) will be available to ensure proper and efficient adaptation of the innovations.

To have the necessary specialized personnel, the S&T infrastructure should be capable of continually identifying their need and training them for the sectors in which research findings can be applied. Likewise, a general level of scientific education among people working in the various sectors is obligatory for proper application of technologies.

The complexity of the elements required to effectuate technological innovations and change are such that a high level of organization and convergence of many different activities are required. Therefore, functional linkages must be established between universities, technical schools and institutes involved in both theoretical and technical research, industry, and other productive agencies.

Along with these linkages must be a highly motivated S&T community of professionals operating under relative freedom, recognition, and recompense.

PROBLEMS ENCOUNTERED IN S&T PLANNING

Many of the difficulties experienced by developing Caribbean countries in achieving S&T competence spring directly or indirectly from a history of colonialistic domination. Colonialistic domination has led to a dependence on the mother country for government, trade, education, and cultural patterns. Essentially, their economies were maintained strictly for advancing the conditions in the mother countries. The colonies were organized to produce agricultural products for the home market with little or no concern for the local people. S&T were developed along these lines in order to exploit a large, unskilled labor force. (12) There was little or no development of indigenous technologies, and the few trained people manned the factories, plantations, and civil service for the plantocracy. There was therefore, no evolution of a scientific tradition within these societies, and this tendency continued during the years of self-government. In the years following full independence, S&T dependence was perpetuated by their relationships with transnational corporations operating out of the former colonial territories. Again, this inhibited the development of indigenous S&T infrastructures and scientific tradition.

With this historical background and social apperception, the construction and implementation of science policies to guide the institutionalization and use of S&T were confronted with other major difficulties, mainly because of insufficient capital, shortage of brainpower, and small internal markets. Some economies are so small that they are unable to support even the rudiments of a S&T system. The fact that countries like Jamaica have to depend on foreign trade for up to 60 percent of the goods and services needed to maintain a viable economy, (13) poses great problems for the effective utilization and development of an indigenous S&T base, and for independent action in technological selection in general.

The major problems confronting small developing countries like Jamaica will be discussed under succeeding headings.

Short Term Necessities Eclipse Long Term Planning

The economic and social conditions in many small developing countries are so desperate that much of the national energies are dissipated on short-term actions to suspend economic and social chaos. These societies are in perpetual crisis, and not inclined to entertain longterm coherent S&T planning.(14) The historical setting and the sociocultural outgrowth of the past has demanded that scientists not only carry out research, but strive to achieve conditions conducive to the conduct of science. They have a clear responsibility to indicate the need to effectively plan for future development while diffusing the imminent crisis situation. Scientists must realize that the custodians of underdeveloped societies often plan from positions of deep insecurity. This is especially true of the so-called democracies in the Third World where political life is short. These politicians, operating in a thankless society, are often totally preoccupied with their own political careers, so they jump from one vote-catching gimmick to another. The concept of real investment in social change, which is the root of all their problems, eludes them. Scientists, therefore, must devise ways of convincing the political planners that their subjective futures can find a purchase in sound S&T policies even though the gains may, in many cases, be protracted. The charlatans of politics, of course, will always be opposed to scientific approaches to social questions. Many of them often cannot weather the scrutiny of reason, honesty, and facts. Their antagonisms and guile should, however, be anticipated, and not be allowed to daunt the forward march of change.

Another problem encountered in planning for S&T in small countries is the paltry material conditions afforded S&T. When S&T policies do not forthrightly remedy these conditions, many workers become frustrated, disillusioned, indifferent, and ultimately recede.

Scientists should then expect shortages of equipment, space, materials, facilities, and emoluments for at least a generation following the formulation and promulgation of science policies. This does not mean that the scientific community should work any slower to achieve the objective ends of proper scientific planning; it simply means that we must be prepared for a long and tedious campaign. In many countries, this is not appreciated, and many scientists and scientific planners become disillusioned and give up the struggle. Each time this happens, it becomes harder for the succeeding groups because the public, by then, will become cynical as to the real objectives of the scientific community.

Government's Responsibility to S&T Goes Unrecognized

Although S&T is a major responsibility of government, it is often not recognized as such. To many leaders, S&T are distractions which keep cropping up when important economic matters are being discussed. This mood predisposes to the conjecture that these matters are best raised and integrated by technicians after the supposedly more important economic questions are settled. Those who argue against this attitude, and insist that sound planning cannot be advanced without relevant technical inputs at the very outset, are often regarded as misguided academics who should be simply humored and ignored. Those among the leaders who are more enlightened may argue that S&T is important, but that small countries with meager resources should not be bothered with things which are outside of their reach. Instead, they should be concerned with the real problems of society, sometimes without identifying specifically what these may be. Others argue that S&T is just too expensive to be taken seriously and should be left for bigger, richer countries. Nevertheless, to maintain a national image, a small financial allocation may be provided for training in science and R&D activities. Occasionally, when it is politically advantageous, there may be ground-breaking and erection of stately buildings for S&T, and for the first years, pronouncements are made of the amounts being given to these institutes. But as soon as the political fervor wanes, they are promptly ignored and fall into disrepair, even before they had a chance to become effective. It is clear, then, that government must be made aware of its responsibility to S&T, and be provided with sufficient information from science advisors to allow the correct decisions to be taken in such a way that support for S&T will be uniform and well conceived. This job falls squarely on the scientific community. The scientific community, for expediency, should not succumb to the urge to coerce support based on false promises, and should not project science larger than it really is. Respect for scientific work will only be engendered when the utility of S&T is clearly demonstrated in tangible terms, along with its limitations and social and economic effects. Politicians must be discouraged from using S&T for short-term political gains which will tend to distort the long-term values of S&T. It is extremely difficult to extricate science from politics in small countries where there is scarce capital. Any move by scientists to treat S&T independently from politics will reduce support by the political directorate. In small countries, it has to be expected that S&T, if properly supported, will become a prime target for politicians. However, if scientists and technologists pay more attetion to the social, economic, and ultimate political aspects of their work, they will be in a better position to convince their political leaders that any attempt to use S&T for purely political ends will end in failure, as these activities have to be based on logic and reason and not political gimmickry. Further, scientists and technologists, if socially aware, can better demonstrate S&T power in increasing productivity and in augmenting the labor force – both of which are politically desirable – irrespective of political ideologies and persuasions.

Similarities in science policies between the two major political systems suggest a certain political neutrality and pragmatism for S&T,(15) but this should not be taken to mean that the uses to which S&T are to be put will not have political implications. All scientists and technologists, therefore, must become more responsive to the society in which they reside and from which they draw support, and strive to become integrated in the political process.

The best way to get politicians to become more aware of their responsibilities is to educate the electorate as to the benefits of S&T. This will not only begin to establish a scientific tradition in these societies, but will stimulate the populous to require more technological information and innovation for their productive processes. If there is sufficient demand by the electorate for S&T, politicians will respond.

Unclear National Future Contributes to Indecisive S&T Planning

Science policies are intended to provide guidelines for the use of science in the affairs of society. If the affairs of society are in disarray, then science policies to service these societies will, undoubtedly, reflect this confusion.(16) Unfortunately, socioeconomic plans for many small countries are ill-defined; consequently, S&T policies are difficult or impossible to construct. It is difficult to formulate a policy in the absence of assumptions of the future trends in a country's plans for production and education. In the absence of a well-conceived national economic and social plan, S&T planning will have no target and, hence, no meaning. The results of S&T activities will mature only after long periods, and, therefore, they have to be projected to meet specific demands of society. A difficulty experienced in small developing countries with agriculturally based economies, such as Jamaica, is the inability to determine the future of their traditional agricultural crops because of a paucity of technological and marketing information. Because of this, it is not clear to the science planner whether to provide for technologies to upgrade the production and use of staple crops, or consider strategies for their substitution. In essence, proper S&T policy planning cannot take place without clearly defined and consistent national planning. Once political directions are set, socioeconomic plans should not be developed without recourse to the most recent S&T information. There must always be an intermeshing and reciprocating system which embodies educational, sectorial, employment, foreign, and science policies.

Shortage of Brainpower

Scientific research and technological development rest fundamentally on creativity, and, consequently, the most important commodity in S&T is brainpower. Without scientists who are energetic, competent, and motivated, a true and vital S&T infrastructure will not flourish or, indeed, will never be initiated. In formulating science policies in countries without a scientific tradition, the notion that the presence of sophisticated equipment and impressive buildings will ultimately produce good S&T has to be put in proper perspective.

Science and technology brainpower is needed initially to formulate S&T policies and later to implement them. The greatest bottleneck to S&T development in small developing countries is paucity of adequate talent. Since the development of appropriate skilled manpower is a long-range proposition, the solution of this problem is difficult and necessitates enlightened national planning.

Historically, in former colonies such as Jamaica, training at the tertiary level was restricted to those professions such as primary and secondary teachers, lawyers, doctors, and civil servants, which serviced the needs of imperial masters, and bore little relevance to actual needs of the local population. After independence, these biases remained as cultural norms, to the extent that institutions of higher learning, which were subsequently built within these regions, still are geared to churn out these types of skilled persons. So, in many developing countries, there is a tremendous imbalance, not only in types of required professional personnel, but also in the ratio of professional to technical workers.(17)

This latter problem has lead to serious obstacles in industrial production. Consequently, in formulating S&T policies, this is one of the first shortcomings which has to be tackled. Unfortunately, without a scientific tradition and an appreciation of the value of S&T, it is difficult to attract young people to technical training, as often this is equated with manual labor, and there is more appeal to office jobs.

The scarcity of proper training facilities, insufficient local demands for indigenous S&T, and improper utilization of skilled personnel usually leads to an unhealthy drain on the few trained personnel. Students who go abroad for training are reluctant to return, and those who return are often employed outside of their field of competence and fall into a nonproductive rut, sometimes without ever utilizing their new-found skills. Others who are willing to remain and make positive contributions are confronted with truncated career opportunities, low prestige, insufficient monetary rewards, a stifling bureaucracy, prevalence of mediocrity and a hidden feeling of inferiority within the scientific community, and insufficient demand for R&D.

Because frustrated senior personnel attracted by lucrative job opportunities abroad often migrate, while there is little or no encouragement for the training of young scientists, the net effect is scarce brainpower. This situation results not only in the loss of investments made in educating the senior scientists, but also removes valuable expertise familiar with local conditions. This brain-drain poses great problems for S&T policies, because it is difficult to plan around few available skills, and because of the uncertainty as to whether the situation will continue to deteriorate or not. One of the features of policies in such countries is to educate the spectrum and number of professionals needed to man satisfactory S&T systems and further to formulate flexible management structures to attract and retain highly skilled individuals regardless of nationality.

Fragmentation of S&T Efforts

As a consequence of insufficient brainpower, marginal governmental support, and an inbred attitude and insularity, a strong sense of protectionism has developed among the scientists in small Caribbean countries. Because of this insecurity, each little group or individual works in isolation to the extent that S&T capabilities are uncoordinated, fragmentary, and fraught with duplication and waste. Each small group often works without the benefit of adequate personnel and equipment which could easily be afforded by collaboration. Policy making is, therefore, viewed with suspicion by those who fear that their small empires will be encroached upon. The level of cooperation from those workers is often much less than desired, and in some cases they may be openly hostile. For these plans to be formulated, and later to succeed, it is necessary to gain the confidence and assistance of these workers by indicating that all will benefit from a process of rationalization. It should be pointed out that while the overall planning of a national science policy should, by its very nature, be centralized as far as possible, to allow for creativity and independence within the confines of coordination and collaboration, research and experimental facilities should be decentralized. Authority to make scientfic decisions should be as close as possible to where action will take place. Science policies in small countries have to grapple with the historical insecurities leading to fragmentation and lack of coordination, while being obliged to enforce the need for a central coordinating center, which will oversee independent units. In other words, the policies will have to discourage independence evolving from affected circumstances while encouraging independence based on scientific creativity.   In small countries, a science center may have to be encouraged to allow for interdisciplinary interaction to achieve the critical mass which is needed to develop and use modern technologies.   The guiding principle should be strong and inspired leadership with maximum freedom for individual scientists. (18)

The attempt to organize these essential units will, as mentioned before, experience some resistance. This resistance may be an out-growth of the feeling that science must resist social and political pressures.   In many small countries, this has resulted in the belief among the citizenry that science has no real productive function. Proper planning will have to take into consideration the need for scientists to interphase with the public as well as the custodians of society to provide the necessary conditions for the development and use of science. A fragmentary and esoteric approach to science has lead to a loss of respect for the scientists in the small countries, and this has to be recaptured by strategies within science policies.

Local Feuds

A most disconcerting problem in the development and implementation of science policies is local quarrels among professionals. Although local squabbles are experienced in all countries, when the few professionals in a small island country become embroiled in personal battles, not only are valuable energies being dissipated, but the necessary coordination and collaboration to ensure a vital indigenous S&T infrastructure are frustrated.

One possible reason for this is the aggressive competition for the meager allotments to S&T, and the fact that incompatible souls find it difficult to avoid each other because of the small size of their community. This type of behavior may also spring from the underutilized energies in these countries. In this climate, science policy endeavors are therefore, stultified because little or no consensus can be achieved by the persons responsible for their formulation. The aura of antagonism can take on a bitter political dimension which completely overshadows objectivity.

Conflicts that are detrimental to S&T planning also arise between the scientists and the civil service administration. Because of the relatively low status of S&T, workers are often subjugated by civil service managers. These managers usually have little concept of the scientists' problems, and often apply strict civil service regulations which are designed for functions very different from creative work. This causes intense irritation and annoyance among scientists and technologists.

A very sore point is the matter of promotion which is handled in a similar manner. In this case, creative and scientific criteria are superseded by the essentials of bureaucratic behavior such as sensitivity, condescension, charm demeanor and dress.

Politics in Science

The lack of sufficient and effective interaction between the scientific community and the political directorate have prevented a realistic science policy. Many scientists consider science above politics, and consequently, are reticent to pursue dialogue with politicians regarding the provision for and utility of S&T. This is further aggravated by the fact that the public is not sufficiently aware of the work being conducted by scientists whom they grudgingly sustain. The consequence of this is that scientists appear to be irrelevant to the national purpose, and a drain on scarce capital. Inferentially, their efforts to formulate policies are regarded as esoteric and purely academic.

It is difficult to extricate politics from S&T in small countries if, for no other reason, because most of the S&T institutions are supported by government since private enterprise is reluctant to do so or finds it uneconomical. In formulating S&T policies, these facts should be appreciated and accommodated. Politicians should be well informed and made sensitive to the delicate intricacies of S&T. Without this sort of rapport, scientific criticism may be interpreted as political dissension. When politics ignominiously supersedes science, then individuals with political sway will gain undue influence in matters which are strictly scientific. However, it has to be fully realized that the translation of science into technologies for social benefits is strictly a political issue. In countries instituting long-overdue change, political leaders will have to consider commitment to change in nominating scarce advisors or leaders. Too often, national programs are regarded as fair game for sabotage by the uninitiated and subversive. The dividing line as to where science leaves off and technology begins is hard to decipher, and indeed there is dynamic social process in operation.

Another problem that faces those who will ensure the implementation aspects of a science plan is that of corruption in government. Since the government is the chief contributor to S&T in small countries, in corrupt or weak administrations, there is a tendency to fill high posts with individuals without the necessary qualifications. This has created, and will continue to create, chaos within S&T systems.

Overplanning

There is a tendency in small countries without a long scientific tradition to consider the planning process as an end in itself. Once a policy is written and scientists can be identified, it is felt that the foundation for S&T is truly laid. Therefore, a lot of time will be spent planning, with little implementation of the plans. The bewildering changes in the socioeconomic environment seem to force certain institutes into perpetual planning with little real research and development. Few programs are meticulously pursued to the very end. Confusion often arises in the attempt to plan for science per se as opposed to efforts to utilize scientific achievements. Further, moralizing and generalizing are mistaken as essentials of a science policy.(19) Executive committees must realize that they cannot produce S&T; at best they can only use it. Overplanning for S&T may also lead to inflexibility which does not allow for feedback and correction of errors. The inability of the plan to undergo necessary change leads to frustration and disillusionment of those involved.

Misguided Education System

The colonial history of small Caribbean islands has left an education system which caters to the affluent elite minority in these societies. The university systems in the Caribbean, especially the British Caribbean, has continued along this same vein – graduating overspecialized individuals with little bearing on the needs of their societies. The few local technologists were trained abroad and had an inclination to return to their training ground as soon as difficulties arose.

Policymakers, therefore, are now faced with the problem of modifying their educational systems to provide not only for top engineers, scientists, and other professionals, but also for badly needed technicians and middle-management personnel which are indispensable for any productive system. The problem of scientific training in these countries starts at the primary levels and later permeates the secondary and higher levels. At all these levels, science is projected as a foreign commodity to be memorized and regurgitated at examinations, instead of as realities which affect everyday life. This leads to the notion that science was the sole prerogative of the well educated, and was primarily an esoteric exercise. Consequently, innovation was not encouraged, and in some instances was openly discouraged, so that by the secondary levels, most students had lost the inquiring spirit which is natural to all human beings. Consequently, to formulate science policies which will ease pressing social problems, involve everyday activities, and organize people, the first step is the education of the public to the true meaning and power of science to dispel the feeling that science is beyond the means of the average person.

Absence of Domestic Demand for S&T

The absence of a scientific tradition, and an appropriate education system to encourage inquiry and experimentation has led to few local scientific discoveries and innovations and an absence of a domestic demand for indigenous technology. The absence of this demand has led some to feel that there is really no need for S&T policies in the Caribbean. Further, it is contended that all our technology needs have been satisfied in the past and this will continue. Therefore, what is required is the purchase of more technolgies to solve all our economic troubles. Science policies in such countries will therefore, have to substantiate the need for local S&T competence by pointing at the necessity for adaptation of technologies, even if we are to be totally dependent on foreign methods and equipment. Needless to say, indigenous efforts will sharpen the local talent and open up wider vistas for the incorporation and use of local and foreign technologies. In many cases, the local demand and concommitant financial support can be governed by selective incentives for the development and use of local S&T and disincentives for import packages.

To create and satisfy the local demand for S&T, science policies in small countries will have to seek ways of linking R&D with productivity on the one hand and education on the other.

Shortage of Foreign Exchange

In the Caribbean basin, except for Venezuela and Trinidad, all states are experiencing grave difficulties with availability of foreign exchange due primarily to spiralling costs of fossil fuels. Jamaica, for example, spends at least 40 percent of its foreign exchange in this manner, so designing a science policy against a background of dwindling financial resources leads to indecisions and lack of confidence, and hence a weak policy. With an unfavorable trade balance and scarce foreign exchange, importation of equipment, international travel, subscription to journals, and purchasing of books are curtailed. Likewise, problems often arise with local customs and the bureaucracy dealing with the issuance of licenses for importation of equipment and supplies. The personnel involved in making these decisions find it extremely difficult to understand the need for scientific equipment against other urgent requests, and will retard applications. Instead of discussing the issues fully, they find it more palatable to frustrate the scientists by delaying tactics which lead to much ill-will on both sides. Since these are aspects of human behavior over which there can be little control, science policymakers inevitably find it difficult to circumvent these issues.

A contributing factor to the foreign exchange crisis is that most of the needs of small developing countries are satisfied by import/export trade. For Jamaica, this represents some 60 percent of our total economy.(20) This fact also places more constraints on planning for S&T development, as this great dependency makes it extremely difficult to institute a decent level of S&T control. Another major decision confronting science policymakers in small developing countries is what portion of the nation's GNP should be allocated to S&T, and further, what fraction of this S&T budget should go to basic or fundamental research. With pressing social problems, such as health and employment, and stringent economic constraints, it is very difficult to ensure that S&T will successfully compete, as decisions are often taken by leaders who are not sufficiently knowledgeable and are insensitive to science and its importance. Policies, therefore, have to be designed to demonstrate that S&T allotments are really investments; indeed they may be considered in many cases as investment in survival. Portions that are slated for basic or fundamental work may have to be regarded as a consumption expenditure.(21) The rationale for justifying public support of these undertakings may consist of treating them as a cultural undertaking or as an overhead to technological development. Whatever way this basic science budget is projected, it has to be comparatively small. If it is too conspicuous, the entire S&T budget may be severely criticized. The problem, therefore, is to strike the happy balance. S&T budgets, in many cases, will have to be supplemented by foreign technical assistance for well-defined projects.

Monitoring of S&T Activities

In an attempt to come to grips with their S&T aspirations, small countries with a colonial background indulge in much planning, little real decision making and even less implementation. Further, where there is implementation, there is no monitoring, no accountability, much failure, and little knowledge gained by these experiences.

Monitoring of science, however, is a difficult affair, although basically there are three criteria which can be used to make an estimate. They are the technological merit of science, the scientific merit, and the social merit.(22) Although the technological implications can, with some difficulty, be assessed, it is even less easy to estimate the scientific and social merits.

Economic theory has not been able to fully characterize R&D as a specific economic activity, primarily because there are no convenient parameters to measure the quality and quantity of science. This problem encourages economic planners to ignore S&T and, in some instances, even ignores the scientists. Science policies, however, must include some evaluation of the impact of its programs, and it must be flexible enough to accommodate change suggested by the monitoring process.

Another aspect of monitoring which science policy must come to grips with is the impact of technologies on the environment. For small island countries, pollution of any kind has devastating and irreversible results which become obvious in a very short time. However, there has been a general neglect of the environment because it is thought by some that small undeveloped countries cannot afford the luxury of environmenental fads which retard technological progress. Any decent science policy, therefore, must indicate that short-term gains at the expense of long-term effects on the environment are eventually uneconomical and foolhardy.

Transfer of Technology

Scientific knowledge is mankind's common heritage evolving over the ages through the creative genius of men from many nationalities and creeds. The fruits of science and technology are, however, zealously guarded by a few in this age who effectively hoard technologies presently in relatively little use but vital to the survival of many. Many small countries are unable, because of size, poverty, and manpower, to generate the minimum of the technologies they require. Needles to say, even those countries with a scientfic tradition cannot develop all the technologies they want, so all countries, to a greater or lesser extent, depend on the transfer of technology. The economies of small developing countries are transfixed without a heavy reliance on foreign technologies. Most of the technology required by small countries is in the hands of private organizations which have much greater allegiance to profit than to any other human cause. Most of these peddlers of technology bargain from extremely strong positions, and, by prohibitive contracts, ungraciously prevent small countries from really acquiring the technolgies which are supposedly to be transferred. Exhorbitant fees are extracted by way of patents, trademarks, licenses, royalties, and other restrictive arrangements. The situation is further compounded by the fact that although small countries normally have some understanding of money and commodity systems, they are usually totally ignorant of the technological markets, and so often make extremely bad purchases. The sellers of technology also pit one small island against another in their attempt to squeeze the maximum returns from the renting of their technolgies. In desperation, small countries get locked into contracts which are not only very costly and marginally economical, but which literally strangle attempts at local indigenous S&T development. Only quality control is allowed or encouraged in many of these contracts. Further, any innovation that perchance is uncovered is considered the property of the licensor while the licensee is dictated to regarding raw materials, consultants, servicing, and markets. Science policies will then have to address these untenable situations without destroying the livelihood of the local entrepreneur, and succumbing to technological rapaciousness. The best way of tackling these problems is for each country to become familiar with its own needs and to strive to gain competence in searching for and selecting technologies on a world-wide basis. Further, no technological packages should be bought without disaggregation by a team of practicing scientists, economic planners, and entrepreneurs who are familiar with local conditions and requirements. Each component of the process should be selected on a basis of social acceptability and economic return and not solely on entrepreneurial convenience.

These policies must, therefore, allow for an efficient information network to provide the necessary data to permit good shopping, proper selection, and suitable adaptation of technologies.

Weak Regional Cooperation

The small size of the various islands in the Caribbean, with their similarity of peoples, climate, and topography, and a commonality of development problems seem to dictate close cooperation and collaboration in the field of science and technology. Unfortunately, a significant level of cooperation is not to be found in the Caribbean. The varied legacies of colonial domination by different national powers have left the islands with a kaleidoscopic assortment of languages, customs, idiosyncracies, dependency patterns, political directions, and a large measure of suspicion, together with an appalling lack of knowledge about their region. However, some collaboration and understanding exist within the English speaking Caribbean.

Indigenous natural resources, although similar in many ways, are diversified enough to allow for beneficial exchanges of raw materials. Needless to say, the size of the Caribbean market available to each country would increase to more respectable levels if an adequate Caribbean market strategy could be devised. S&T collaboration would not only increase the number of scientists, technicians, and materials to tackle any one problem of mutual concern, but would also enable problems which require large outlays, beyond the reach of any one small country, to be broached. The rewards to be gained by establishing systems for S&T information exchanges, with full cataloguing of the traditional technologies which abound in the region, are incalculable. Further, the extremely small size of many of these islands precludes any decent S&T system. Consequently, regional cooperation would benefit them immeasurably. At best they will have to rely on S&T extension services from larger neighboring territories for their basic needs. Science policies in these islands will, therefore, have to grapple with establishing lines of communication, and technological cooperation to enable the best use of the natural and human resources and experiences in the region.

NOTES

(1) Population, Per Capita, product and growth rates. World Bank Atlas, Washington, D.C., 1976.

(2) World Economic and Social Indicators. World Bank Report No. 700/78/01, Washington, D.C., 1978.

(3) Suggestions for the preparation of National Papers, United Nations Conference on Science and Technology for Development, COSTED, Madras, India, 1977.

(4) Science, Technology and Developing Countries UN Document 77-75589, New York, 1977.

(5) International Development Strategy: Action Programme of the Second United Nations Development Decade, UN Publication No. E.71.11.A.2, 1971.

(6) B.V. Rangarao "Science Policy: Role of Academic Societies," Science and Culture 43 (1977): 194-200.

(7) Promotion of the Formulation and Application of Policies and Improvement of Planning and Financing in the Field of Science and Technology." Mid-Term Plan (1977-1982) UNESCO. 196/4, Paris, 1977, p. 111.

(8) J.A. Stratton, "Changing Role of Science and Technology," Nature 203 (1964): 455-57.

(9) R. Dubos, Reason Awake, Science for Man, (New York: Columbia University Press, 1970).

(10) Science and Technology in Asian Development, UNESCO, SC. 61/D.69/A, Switzerland, 1970.

(11) Dubos, op. cit.

(12) N. Girvan, Caribbean Technology Policy Studies Project, General Study: Preliminary Report, Institute of Development Studies, University of Guyana and Institute of Social and Economic Research, University of the West Indies, 1977.

(13) External Trade, Department of Statistics, Kingston, Jamaica, 1976.

(14) J.A. Sabato, "Quantity versus Quality in Scientific Research (1): The Special Case of Developing Countries." Impact of Science on Society 5 (1972): 29-41.

(15) S.G. Allende, "Science in Chile's Development Programme," Impact of Science on Society 22 (1972): 29-41.

(16) J.J. Salomon, "A Science Policy for the 1970's," OCED Observer, No. 53, 1971.

(17) Survey of Jamaica Science and Technology Research Institutions and Preliminary Outline of Development Need, The National Planning Agency, Kingston, Jamaica, 1975.

(18) J.J. Moravscik, Science Development. The Building of Science in Less Developed Countries, (Bloomington, Indiana: PASITAM 1975).

(19) National Academy of Sciences. Applied Science and Technological Progress, Report to the Committee on Science and Astronautics, Washington, D.C.

(20) External Trade, Department of Statistics, Kingston, Jamaica, 1976.

(21) Science and Technology in Asian Development, op. cit.

(22) Dubos, op. cit.

16 Science and Technology Planning Problems in a Large Circum-Caribbean Country (Mexico)

Miguel S. Wionczek

More than half a century after the revolution, Mexico still belongs to the underdeveloped world, with the majority of its rapidly growing population existing under very precarious conditions. For years, historians, sociologists, political scientists, and economists have been trying to explain why the country has continued to be underdeveloped. Most current explanations are, however, at best incomplete and at worst based on unverified assumptions.

For a long time, it was claimed, both inside of the country and abroad, that the slowness of socially acceptable development in Mexico was largely due to limited natural resources, a relatively small population dispersed over a very extensive territory, and a per capita income too low to generate sufficient savings to finance growth. However, more and more evidence is available that Mexico is not at all poor in natural resources, but rather that only a minute part of these resources has been surveyed, and an even smaller portion actually exploited. The recent discovery of new oil resources offers evidence in that respect. Nor can one speak of a shortage of human resources in a country with one of the highest demographic growth rates in the world (3.2-3.5 percent a year) and with a labor force expanding by about one million people a year. Moreover, with the present annual per capita income of about $1,000 and with a high concentration of income, it cannot be assumed that Mexico's potential savings are low. Thus, it would appear that the explanation for the persistence of Mexico's economic and social underdevelopment must be sought elsewhere. Without underestimating the difficulties arising from the nature of international economic relations between developed and developing countries, it is possible to maintain that among the most important causes of Mexico's relatively low development level are the absence of political modernization, the severely deficient social organization, the very poorly designed educational system, and scientific and technological backwardness. Together, these factors result in inadequate and inefficient management of the country's major problems.

MEXICO'S DEPENDENCY

The weakness of domestic scientific and technological activities reflects the backwardness and dependency of the Mexican economy. Mexico's rapid quantitative economic expansion, accelerated industrialization, and scientific and technological development date only from the 1930s. It was at that time, less than 50 years ago, when, by putting university education on the list of priorities together with official support offered industrialization, conditions were created propitious for some scientific research, however unaccompanied, by similar efforts in the field of technology .

As a result, while scientific activities in the major universities have registered some progress in recent decades, the advance of applied research and development was retarded by conditions arising from the dependent and imitative industrialization process. In response to local technological backwardness, technology embodied in machinery, or available through licensing of industrial know-how, was massively imported, even though these imports very often did not correspond to the needs of a country short of capital and faced with an abundance of unskilled labor. Not only were industry demands for imports of technology and services linked to the production structure, serving mainly the urban groups whose preferences were formed by consumption patterns of high income countries, but often imported technology required intermediate inputs not produced in Mexico, and at times even hindered the exploitation of some widely available renewable resources. The nature of its technological links with the outside world changed little after World War II, when Mexico's industrial structure expanded with the establishment of many intermediate goods industries and the appearance of an incipient capital goods sector.

Thus, in most recent times, Mexico's scientific and, in particular, technological dependency upon the outside world increased instead of diminishing. Isolated attempts to rectify the situation by fostering R&D activities at the universities, and in the public sector, were not accompanied by the elaboration of a national policy for science and technology until very recently. Only with the setting up of the National Council for Science and Technology (CONACYT) at the beginning of the 1970s, was the need for a science and technology policy perceived in some quarters. This coincided with the worldwide recognition of the interrelationship between scientific and technological effort, and the patterns and pace of economic and social growth. Consequently, CONACYT, in close cooperation with the scientific community, leading universities, and representatives of public and private sectors, formulated, in 1975/76, the first Science and Technology Plan, covering basically the period 1976-1982, but within the framework of a longer term R&D strategy extending to the end of the present century. The contents of this plan, which, for reasons to be explained later, has been largely abandoned with the change of administration at the end of 1976, will be reviewed here at some length, in order to throw some light upon the problems arising from the scientific and technological backwardness of practically all developing countries.

PRESENT STATE OF SCIENCE AND TECHNOLOGY IN MEXICO

A detailed survey of scientific and technological activities undertaken in 1973-75 revealed that Mexico's R&D system, although having accelerated its growth in the present decade, faced formidable difficulties, many of which are common to most Latin American S&T systems.

1. Mexico depends to an exaggerated degree on the development of science and technology in more advanced countries, thus limiting its output in many cases to purely imitative quasi-research activities in fields in which serious local R&D is badly needed, if only because many problems arising in the context of underdevelopment are different from those facing developed societies.

2. Financial resources available domestically for R&D are not only inadequate in comparison with those provided by industrialized countries, but also as compared with R&D expenditures in some countries with a similar development level, such as larger Latin American republics.

3. The science and technology system counts neither the quantity nor the quality of human resources required, both in absolute terms and in comparison with many other countries of a similar level of development.

4. Geographic and institutional concentration of science and technology institutions is excessive. In 1973, research institutions located in or around Mexico City accounted for more than 80 percent of total expenditures and personnel, and five institutions spent 45 percent of the national R&D budget.

5. The functional distribution of R&D expenditures is deficient. Almost 70 percent of financial resources is spent on salaries and wages, while less than 15 percent is available for purchasing equipment indispensable for serious research.

6. Most R&D institutions face a critical lack of researchers. Only 3.5 percent of the total of 400 existing research entities employ more than 20 people each, the minimum needed for relevant research in most fields.

7. The development of science and technology is highly unbalanced sectorally and by disciplines, with consequent neglect of very important areas of research. Resources for basic research are extremely scarce, and applied research and development is concentrated in those few sectors where the presence of the state is particularly great. Petroleum and energy, modern agriculture, medicine and health, and the intermediate goods industries absorb half of the financial resources available. Even

in these fields, research is neither sufficient nor adequate to satisfy the country's specific needs for scientific and technological knowledge, Furthermore, R&D is neglected in such areas of importance as subsistence agriculture, nonrenewable resources, capital goods, transport and communications, and urban development and housing.

8. There are no permanent links between the R&D effort and the educational and productive systems. Moreover, the structure of the science and technology system fosters a divorce between R&D and dynamic and technically complex productive activities. The weakness of technical diffusion and extension services obstructs the transmission of knowledge to the productive system, especially in noncommercial agriculture and the consumer goods industries.

The confrontation of existing situation with the country's probable R&D needs indicates that the science and technology system should expand its sphere of action considerably, in both quantitative and qualitative terms. It indicates, moreover, that the future growth of the system and the relevance of R&D depends largely on the formulation of a long-term overall development program, badly missing in Mexico for sui generis political reasons. In other words, the future of science and technology in Mexico seems to be contingent not only on the increase in financial resources for R&D and on the more rapid training of human resources, as is traditionally claimed, but also on the integration of these efforts into a general planning framework. Given the nature of scientific and technological activities, the characteristics of the Mexican economic model and the official ideology, such planning cannot be anything but indicative and participative. In respect to science and technology, such planning should not only foster the expansion of R&D, but aim at overcoming a practically nonexistent demand for domestically produced scientific and technological knowledge. The work on the Mexican S&T plan demonstrated beyond doubt that in the developing countries the mere supply of know-how does not create the demand for it.

THE MEXICAN SCIENCE AND TECHNOLOGY PLAN

The Mexican Science and Technology Plan was based on two premises: first, that the importance of science and technology for socioeconomic development makes its long-term planning an urgent necessity for any country;   second, that the need for scientific and technological planning, while safeguarding freedom of research, is even more pressing in countries like Mexico, owing to the persistence of underdevelopment, the relative scarcity of the government's financial resources, and the magnitude of the still unsatisfied basic needs of the majority of the population.

The main objectives of the Science and Technology Plan were

defined as "non-imitative scientific development, cultural autonomy, and technological self-determination." Scientific development should be understood here as the creation of a capacity for research in the exact, natural, and social sciences which would enable the scientific community to fulfill its social function and, at the same time, to participate meaningfully in the process of international scientific advancement. Cultural autonomy is an objective related to safeguarding certain social values which are being lost in the process of industrialization in many developing countries. Technological self-determination is defined as the construction of a domestic capacity that would permit the demand for technology to be reoriented progressively, to the extent needed, toward local sources of technical knowledge that would rationalize purchases of foreign technology and help to assimilate and to adapt imported know-how, using it as the basis for internal generation of technology .

The scientific and technological policy required to attain these long-run objectives postulates that:

1. Science and technology policies must be thoroughly integrated with the country's general development policy.

2. The model of scientific and technological advancement must be adapted to the country's long-term social and economic objectives.

3. The adoption of an autonomous model for the advancement of science and technology in no way implies abandoning the selective use of externally generated scientific and technological knowledge.

4. In the absence of a well-integrated science and technology system, the task of overcoming the present state of scientific backwardness requires a joint sustained effort on the part of the government, R&D entities, higher learning institutions, and the productive system.

5. Scientific and technological advancement demands a favorable environment that acknowledges its social value, and particularly its contribution to the achievement of long-run national objectives.

6. A degree of excellence must be attained in certain scientific fields, little explored until now, and developed elsewhere, but of great relevance for the solution of the problems of underdevelopment.

7. Meaningful technological advance requires parallel and simultaneous activity on several fronts: in selected areas of conventional R&D practiced in the advanced countries, in appropriate "primitive" technologies of local origin, and in some specific fields where the dynamics of advanced R&D offer hope for

major, and early, technological breakthroughs of social importance.

8. The science and technology system must have close links with
the educational system and the economy.

Considering the state of underdevelopment prevailing in most R&D
supportive activities, and the urgent need to transmit R&D results to
the society for educational and productive purposes, the plan dedicated
considerable attention to the problems facing the scientific and
technological infrastructure, including training of high-level human
resources;   diffusion and distribution of knowledge, information, and
statistics;   engineering and consultancy services;   production and
maintenance of scientific equipment and instruments; and international
scientific and technological cooperation. For each of these components
of the system's infrastructure and its external linkages, the plan
attempted to determine both long-term objectives and medium-term
policy guidelines.

In respect to R&D itself, the Plan defined objectives and set
guidelines for problem-oriented research activities, in both the exact
and natural sciences, and in social disciplines. In the field of applied
R&D, it developed technological policy guidelines for nutrition, agriculture and forestry, communications, urban development, industry, energy, renewable resources, construction and housing, medicine and health,
educational techniques, and natural phenomena research. The plan
emphasized that scientific advancement must be based not only on the
recognition of university autonomy, but on the state's commitment to
guarantee the academic and research freedom necessary to foster
scientific creativity. The specific S&T policy guidelines were expected
to be translated later into institutional and sectoral programs, with the
aid of some relatively simple interinstitutional mechanisms designed for
that end. Just as in the formulation of the Science and Technology Plan
itself, these mechanisms for programing and implementation were to be
indicative and participative.

The plan postulate originally that expenditures on science and
technology in Mexico should reach, by 1982, a total of 16,200 million
pesos ($1,300 million at 1975 prices), i.e. almost three times the 1976
spending and slightly more than 1 percent of GNP. The outlay for
research and development alone was to rise from 3,090 million pesos
($250 million) in 1976, to 9,200 million pesos ($735 million at 1975
prices) by 1982. A slight reduction in the government's share of total
expenditures, from its current level of 80 percent to 75 percent, was
assumed, along with the proportionate increase in private sector
spending. External financial aid was expected to be marginal. Furthermore, the plan set up medium-term targets for R&D support activities,
and outlined a package of science and technology policy instruments
aimed at increasing R&D productivity. Finally, it proposed the setting
up of mechanisms evaluating the progress of the plan's implementation.

The plan proposed that S&T policymaking be institutionalized
through the permanent National Scientific and Technological Planning

Commission, to be composed of high level representatives of the federal government, decentralized agencies and public enterprises, higher education institutions, and science and technology users in the productive sector. The National Commission would first coordinate and guide the preparation and the periodic revision of S&T policies, as well as of successive plans; and, second, guarantee a serious involvement in these tasks by the government itself, higher education institutions, the private sector, and the scientific and technological community.

It was proposed that the Planning Commission would be supported by the Inter-Institutional Science and Technology Committee, composed of the CONACYT and various key ministries in charge of allocating and controlling the use of financial resources. The main function of the committee was the integration of the annual federal budgets for science and technology into the framework of longer-term financial targets and R&D guidelines, set by successive science and technology plans. Together, the Planning Commission and the Inter-Institutional Committee were to form the basis of the permanent planning mechanism for the science and technology system, planning closely interlinked with the socioeconomic development strategy to be established by the federal executive.

PERMANENT PLANNING PROCESS

The permanent planning process was to be comprised of four phases:

1. The formulation of a strategy for science and technology development within the country, with a long-term perspective (20-25 years);

2. definition of a medium-term (10 years) science and technology policy;

3. formulation of successive indicative six-year science and technology plans; and

4. elaboration of institutional and sectoral general R&D and R&D-support programs for the duration of each plan.

Given the necessarily long period required for scientific and technological efforts to show tangible results, and considering its current underdevelopment in Mexico, the authors of the first plan assumed that unless science and technology planning is based upon a long-term strategy, it runs the risk of being reduced to an insignificant exercise with extremely limited effects. While the planning might increase the volume and the quality of local R&D, in the absence of a long-term strategy it could not assure the relevance of such R&D from the viewpoint of the country's needs, or the establishment of permanent links between science and technology production and the educational and the productive systems. The need to define science and technology policy for 10-year periods only, and to redefine them periodically, was determined by the nature of scientific progress and technological change. Progress and change the world over is so rapid that science and technology objectives and policy guidelines must be revised from time to time in light of these developments. The need for formulation of successive six-year plans arose from the necessity for any realistic planning to take into account Mexico's political and administrative cycle. Also, it was considered important that institutional and sectoral R&D programs be worked out by the scientific and technological community itself, so that R&D operative continuity would be assured, its productivity improved, and the existing infrastructure facilities and financial and human resources used more rationally.

The Science and Technology Plan wanted to leave scientific and technological institutions free to define their short and medium term work programs, within the framework of policy guidelines and priorities established by consensus by the plan itself. Institutional programs were expected not only to contain eventual programs for research and related activities for the duration of each plan, but also to outline in a general and preliminary way problem-oriented high priority research areas for longer periods.

Institutional programs were to be coordinated by sectors, not to meet formal requirements of planning, but because the scientific and technological policy guidelines presented in the plan strongly indicated that the Mexican science and technology system will have to undertake in the coming years many multidisciplinary R&D activities that exceed the capabilities of any single institution. Linking institutional and sectoral programing with the formulation of the annual federal budget for science and technology was meant to enable a more efficient allocation of government financial resources, and to insure their use in accordance with the policy guidelines of the plan.

Although the very nature of S&T planning requires it to be indicative, because the R&D activities do not lend themselves to any other approach, various groups of participants in the science and technology system were to have different degrees of commitment to the implementation of the plan's overall and sectoral guidelines.

Thus, the Science and Technology Plan was to be mandatory for the National Science and Technology Council, which was expected to act as a technical advisor to scientific and technological institutions at their request, and at the same time design its own R&D support program for the duration of the plan in those fields not covered by other institutions because of insufficient resources or the lack of infrastructure. Thus, within the framework of global science and technology planning, CONACYT was to continue to foster certain activities which R&D institutions could not undertake by themselves. The council's priority activities were to be: training of high level personnel; setting up research centers in sectors lacking adequate institutional structure; and selectively promoting mechanisms needed to link science and technology with the educational and the productive systems.

Government S&T research institutions, centers, and units were also asked to accept a strong commitment to the plan's objectives, targets, and policy guidelines. The argument in favor of this commitment was rather simple. Not only did the plan correspond to a federal program, but the federal government was the direct source of funds for R&D activities undertaken within the public sector. It was clearly understood that the support for the plan at the highest political level was a necessary condition for getting some 300 government-owned research institutions of all possible sizes to participate in the plan's implementation.

PRODUCTIVITY OF SCIENCE AND TECHNOLOGY EXPENDITURE

Although the state and its enterprises account for some 60 percent of national expenditures in R&D, in many cases the productivity of the government's direct expenditures in S&T research is very low, due mainly, though not exclusively, to two factors. First, while there are a few large research centers, such as the National Institute for Agricultural Research and the Mexican Petroleum Institute, whose contributions to national science and technology are unquestionably outstanding, there is in Mexico an excessive number of small research units which, because of size, budget, human resources, and bureaucratic problems, cannot be expected to make any relevant advances in the field. (There are in Mexico some 300 public sector R&D units each with three or less researchers.) Second, many of the small units rarely undertake formal research programs and suffer from a lack of coordination, either within the institution to which they are attached, or with other institutions carrying out similar research. If these small units are to increase their productivity and undertake more relevant research, the government would have to involve itself specifically in the implementation of the Science and Technology Plan, while creating the conditions which would permit these units to establish their programs with a view towards participating in sectoral R&D planning.

Fortunately, all these deficiencies are less frequent in the majority of R&D centers belonging to the major higher education institutions. Consequently, and also in response to the concept of university autonomy, the Science and Technology Plan was to be indicative for both the university and the autonomous research centers. Moreover, it is worth keeping in mind that the plan's broad research policy guidelines were drawn up and adopted by consensus with broad participation by the scientific and technological community, the majority of whom work at the higher learning institutions. Consequently, the planners expected that it would be relatively easy for major university R&D centers to accept the guidelines of the plan and to formulate institutional research programs within the plan's general framework.

PRIVATE SECTOR PARTICIPATION

There still remains the question of the private sector's participation in research and development. In view of the general lack of interest in R&D by business firms established in Mexico, the strong particpation of foreign capital in the technologically dynamic sectors, and the very marked preference for foreign technology by both foreign and local enterprises, it was considered highly unlikely that the private sector would substantially modify its technological conduct in the shortrun. Consequently the plan contended that the government must implement fiscal, financial, and other mechanisms to stimulate private companies to develop their technological capacity, to use research originating within the country, and to increase their participation in the domestic science and technology effort. It was also considered necessary to design instruments which would encourage the large foreign-owned companies to adapt their technology to local conditions and requirements, since Mexico could not remain, from the viewpoint of national interest, in a complete and perpetual state of dependency on foreign technology.

LIMITATIONS OF SCIENCE AND TECHNOLOGY PLANNING

Nevertheless, the plan assumed specifically that even the best planning and implementation of scientific and technological programs at the institutional, sectoral, and national levels could not guarantee the contribution of science and technology to Mexico's socioeconomic development. Contrary to the contention of some recent thinking in the industrialized world, the main limitation arises from the fact that science and technology by itself cannot solve the major underdevelopment problems, although it may provide some elements essential to a solution.

In other words, as stated earlier, in order to take full advantage of the science and technology potential to help fulfill overall national goals, S&T policies had to be coordinated with the long-term development policies. In operational terms, this would amount to the introduction of a number of direct S&T policy instruments, and corresponding readjustments in the existing economic policies that directly affect the functioning and development of S&T activities.

The impact of industrialization policies on science and technology emerged as one of the particularly urgent problems in this area. These policies, together with the supporting fiscal, monetary, and foreign trade policies have not, in Mexico, as yet taken adequate account of the need to accelerate scientific development or to promote technological self-determination. In fact, many of these policies have had a negative impact on science and technology objectives. Moreover, the divorce between important economic policy instruments in force and the proposed science and technology policy instruments was observed. This divorce originated in the fact that most economic policies had been elaborated at a time when the relationship between scientific and technological activities, on the one hand, and socio-economic development, on the other, was not clearly understood. Consequently, the science and technology planners discovered that, like many other developing countries, Mexico was facing the difficult and complicated task of integrating these two policy areas into a coherent whole.

Unfortunately, with the change of the administration such integration proved impossible.

THE DEMISE OF THE PLAN

From the perspective of early 1978, about one-and-a-half years after the official release of the first Science and Technology Plan in Mexico, it seems clear that the plan is not going to be implemented during the present administration that started in December 1976. The reasons for the early demise of the plan are manifold. While some observers believe that the planning exercise was badly timed in political terms – the plan was made public at the very end of a six-year political and administrative cycle in Mexico – it is difficult to believe that such factors bear exclusive responsibility for the abandonment of the first serious attempt to engage in science and technology planning in Mexico. In retrospect, it seems much more probable that the authors of the plan overestimated the interest of the Mexican political system in science and technology, at the same time underestimating the dead weight of federal bureaucracy and parochial interests of the small national scientific and technological communities. Moreover, the publication of the plan, which received formal approval of both the outgoing and the incoming presidents, coincided with the most serious political and economic crisis of postwar Mexico. Under these conditions, the scientific and technological problems lost importance relative to the many other pressing problems that the government and public sector had to face, particularly because its authors insisted on the long-term nature of any effort in their field, and was unable to offer politically attractive solutions to the country's short-run problems.

Independent, however, of the lack of Mexico's success with the implementation of its first Science and Technology Plan, that exercise in planning offers many lessons to other underdeveloped countries. Since these countries will have to engage in science and technology planning one day if they want somehow to diminish their dependency upon the outside world, the summary presentation of major lessons may be in order.

LESSONS FOR UNDERDEVELOPED COUNTRIES

The first lesson amounts to the need to recognize that science and technology problems in the context of general underdevelopment differ basically from those encountered by science and technology in the advanced world. Thus, the advancement in this field in the underdeveloped part of the international economy can hardly be achieved by the same methods successfully applied in the world's industrial centers. Since scientific and technological backwardness is part and parcel of overall underdevelopment, science and technology policies must be integrated into the general development policy framework. The absence of such a framework limits severely the relevance of any attempt to build up domestic scientific and technological capacity.

The second lesson is that one of the major obstacles for the advancement of science and technology in a country like Mexico originates not only in political factors, but in the divorce between local R&D activities and the educational and production systems. Consequently, whatever knowledge is produced domestically is used neither for the improvement of the quality of education nor for production purposes. Mexican experience strongly suggests that the supply of internally-produced scientific knowledge and technical know-how does not create an automatic demand for them, because historically the demand has been directed to the outside. Consequently, the advancement of science and technology in an underdeveloped country depends more upon the effort to establish links between the R&D system and the education system and the economy, than upon the simple increase in human and financial resources allocated for R&D. The acceptance of this proposition will make it easier to understand why a science and technology strategy proposed for the underdeveloped countries by the advanced world, and postulating the establishment of local modern scientific institutes more or less at random, while leaving applied R&D effort to traditional international transfer mechanisms, just cannot work. In the absence of demand for their output, modern scientific institutes set up in the underdeveloped societies wither away and become focal points of the brain drain. At the same time, the dependence on traditional technology-transfer mechanisms leads to the emergence of advanced technology enclaves that perpetuate themselves in the context of general technological backwardness. The question here is not whether such technology-transfers, for example, through foreign owned enterprises, are of any use in the absolute sense. They may be useful or useless depending on the presence or the absence of other vehicles for transfer and propagation of technical know-how in an underdeveloped society. But only the technological strategy designed to establish permanent lines between technological imports and domestic R&D systems, on the one hand, and the local R&D output and the educational and productive systems, on the other, can assure any meaningful long-run technological modernization in a backward country.

The third lesson, originating from the two previous ones, can be summarized as follows: a domestic science and technology system in an underdeveloped country must be defined not just as the sum of local R&D producing entities, but as the universe of all units dedicated to R&D, to R&D supporting activities, and to intermediation between the R&D institutions and higher learning bodies, as well as with the productive enterprises. The intermediation is not one-way only – from those who produce knowledge to those who use it – but it should be visualized as a sort of sending-and-receiving triangular relationship. If science and technology policymakers in the advanced countries seem to forget this, it is because they lack historical perspective. In all the advanced societies the sort of triangular relationship between science and technology and education and production, absent in the backward societies, was built up slowly – and one has to admit, without any sort of planning – over the last two centuries. This statement also covers socialist societies contrary to widespread beliefs. They were not scientifically or technologically backward in their presocialist times, particularly when compared with the majority of underdeveloped countries as we know them today. The Soviet Union before 1917, Poland before 1945, and China before 1948 were quite advanced in many respects in comparison with most of Latin America, Africa, and Asia of the mid-Twentieth century. Thus, if one wants to advance science and technology in the underdeveloped world, one faces the difficult task of devising policy instruments affecting the broad R&D system as defined above, and at the same time revising educational and economic policies in light of scientific and technological effort.

The fourth major lesson of the Mexican exercise may, perhaps, be that we know precious little about the intrarelations, particularly in the context of underdevelopment, within the continuum known as R&D. The simplistic proposition that every country needs to support, in a similar way, all parts of that continuum (because allegedly pure science is needed to prepare the ground for applied scientific effort needed, in turn, for technological development) is open to many criticisms on logical, structural, and historical grounds. Only by accepting our ignorance in respect to intrarelationships within R&D, agreeing that social functions of different parts of the R&D continuum vary considerably, and relating the production of knowledge to some overall view of long-term social, economic, and national objectives in a given society is it possible to arrive at a broad vision of national science and technology strategy for an underdeveloped country or region.

The final major point worth making is that science and technology policy problems cannot be meaningfully handled just by scientists and technologists, if only because science and technology is not a specialized sector, but affects every phase of social, economic, cultural, and even political life. If we accept, furthermore, a proposition that science and technology is not socially neutral, we may arrive at the conclusion that planning scientific and technological endeavors is a very complicated matter in which all available wise men from different walks of life, including wise politicians if available, should perhaps participate. This may be particularly true in the context of underdevelopment, where it may well be that scientific and technological elites, while highly education are, in other respects, as backward as the societies in which they function.

V

The U.N. Conference on Science and Technology for Development

17 Will the UNCSTD Break Through the Action Barrier?

Michael J. Moravcsik

Meetings on international aspects of science and technology, like meetings on any other subject, can operate on three distinct levels.

The first of these could be called the platitude level. Here the meeting is filled with colorful rhetoric appealing to "new economic order," "international justice," "spirit of cooperation," "eradication of hunger," and other noble-sounding phrases, some quite undefined, others utterly general and vague. At this level, agreement among all participants can be guaranteed even in advance since, after all, few countries would go on public record opposing virtue and advocating sin. At the same time, this level is completely nonfunctional in that not only does it not lead to action, but it does not even contribute to the conceptual clarification of issues.

The second level might be described as one creating a framework. It involves discussions about subjects like the relative merits of science and technology, the advantage and disadvantage of traditional technology transfer, appropriate technologies, the economic and moral bases of patent systems, or the relative advantages of research councils vs. ministries of science and technology. This, from a conceptual point of view, is a much more rewarding stage than the first one. At the same time, however, it is at this level that great difficulties arise, because different people and different countries might have quite different conceptions of what these frameworks for international science and technology ought to be like. Contrasts between views occur not only along developed vs. developing countries, or countries with private enterprise vs. countries with government monopolies, or Europe vs. Asia, but also economists vs. scientists, and politicians vs. development specialists. Since the development of science and technology is far from being a science at the present time, such discussions seem to have a perpetual character in which insights may be gained, but resolution and consensus are not to be expected.

Finally, the third level is that of proposals for specific action programs. On this level, the concern focuses on a specific deficiency that exists today, and mechanisms are suggested and debated on how to remedy or alleviate that particular problem.

It might be thought that the third level could not be reached without having resolved all or most of the problems encountered on the second level. This, however, is not so. It is striking how people with radically different conceptual frameworks, or even different long-term objectives, can easily agree on a certain specific deficiency being vexing, and on certain modes of action being helpful in relieving this deficiency. In fact, one could state without much exaggeration that it is on this pragmatic level that most human progress is achieved.

It is, indeed, most salutary that progress on the third level can be made without having passed the second level successfully, because if this were not so, the second level might very well represent an insurmountable barrier to any action.

Past international meetings on science and technology generally failed to reach the third level. After dwelling extensively on the first level, they entered the second level only to become embroiled in acrimonious debates on ideological or theoretical issues and views. In order to save the conference, the proceedings were then guided back onto the first level where there was sufficient safe ground to create a mellifluous closing statement to which everybody could subscribe. The fact remains, however, that the action barrier was not broken and, hence, very little operational benefit resulted from the meeting.

Preparations for the 1979 United Nations Conference on Science and Technology for Development (UNCSTD) have so far taken place partly on the first level (and these can safely be ignored from any practical point of view), and partly on the second level. As mentioned earlier, discussions and contributions on the second level have some intrinsic value. It is, however, crucial for the success of UNCSTD that these activities on the second level be definitely subordinated to rigorous preoccupation with specific issues on the third level.

There are huge areas of deficiencies in the present international structure of science and technology which, in virtually everybody's opinion, should be eliminated. The scientific and technological communities in the developing countries are small, fragile, and isolated from the mainstream of the growth of knowledge and know-how. Educational systems in developing countries in the areas of science and technology are insufficiently staffed, poorly equipped, and too small in size. The outlook of science and technology in the developing countries is limited to too few people, and the population as a whole has not been affected by it.

These thoughts are, of course, very broad by themselves, but within them there are a multitude of specific problems awaiting specific solutions. Just to mention one example, mechanisms to channel unused scientific and technological journals to libraries in developing countries are not hard to find, if enough people concentrate on the problem and are willing to make small adjustments in existing practices.

It would seem, therefore, that the most useful way of preparing for UNCSTD would be for groups and institutions to have work sessions at which such specific mechanisms and their logistic and financial details are formulated, published, and then forwarded to UNCSTD. Correspondingly, UNCSTD itself should also put aside at least some time for the discussion and agreement on some of these proposals. It is most likely that such proposals will constitute the only operationally significant result of the huge effort that will have gone into the organization of this giant meeting. In fact, even if UNCSTD itself never reaches the third level, agencies, professional societies, groups, and individuals attending or following the happenings at UNCSTD might acquire some of these proposals and turn them into action on their own. This is, in fact, the main hope for UNCSTD anyway: It might serve as source material for scores of national, international, governmental, and private organizations and individuals for enhanced international scientific and technological cooperation.

We must, therefore, make sure that this time the second level does not act as an insurmountable obstacle to action, but that a strategy is devised to break through this action barrier and thereby open up the huge potential of benefits that could from the proper channeling of all the efforts and activity that surround UNCSTD. Everyone can contribute at least a small share toward this penetration of the barrier, and everyone must do so.

18 Predictable Failures of the UNCSTD

Stevan Dedijer

The prediction in the title of this paper was inspired by the basic document of January 1975 preparing the United Nations Conference on Science and Technology for Development. The panel of experts set up by United Nations Economic and Social Council produced its first draft in November 1974 and expected to finalize it by May 1975. It is my aim to disprove the prediction by pointing out a number of built-in failure mechanisms in the January 1975 panel proposal, which defines the approach, objectives, topics, participation, and agenda of the conference.

The first such mechanism is that the panel does not practice what it preaches, "universal involvement" in the preparation of the conference. The panel has followed the traditional United Nations committee work ritual and routine. By keeping its first proposal "restricted" and by not exposing its earliest phase of work to public scrutiny of all the conflicting approaches to the problem existing now in the world science and technology policy community, the panel set very narrow limits to the participation, approaches, objectives, etc. for the conference. It also severely limited the time exposure for the examination of the proposal before the General Assembly acted on it in the fall of 1975. I hope to implement the 'involvement' goal of the panel by initiating 'prematurely' the debate on the proposal.

In stating that there has been a "very limited progress achieved in the application of science and technology to problems of development" the panel is echoing a widespread belief in the world S&T policy community. One finds it, among a hundred other places, in Graham Jones' 1971 book, "Science and Technology for Developing Countries," written for the International Council of Scientific Unions (ICSU); in its 1974 counterpart prepared by the Soviet Academy of Sciences; in the 1974 memorandum circulated at the diplomatic level by the Algeria Group of 77 committee on S&T; and in the OECD October 1974 invitation to a conference on the same subject. "The application of science and technology to the developing countries does not appear to have made great headway." In its list of four proposed conference topics, under Topic I, the panel asks the central question related to this belief: "Why has the contribution of science and technology to development of developing countries not been more effective?" The question turns out to be pure rhetoric, for there is no further mention of it in the detailed outline of Topic I, which deals with an entirely different subject, nor is it mentioned anywhere else in the rest of the document.

The failure to base the whole proposal on the thorough exploration of this fundamental question, to devise tools for its study and to draw consequences from its results, constitutes in my opinion the second, and perhaps the central, of the built-in failure mechanisms of the conference. For the failure to build the conference around this question is an invitation to everyone else to be silent about it.

This can best be appreciated in view of the enormous effort − not estimated or even mentioned in the proposal − invested in the problem during the past 15 years by such components of the world S&T community as the national S&T policy bodies, including aid-receiving and aid-giving agencies: regional entities like the Andean Pact, COMECON, and the European Economic Community;   United Nations regional bodies, specialized committees and agencies;   international organizations from OECD, Algeria Group of 77, to ICSU, World Federation of Scientific Workers;   and hundreds of academic institutions and individuals.

One illustrative, though imperfect, measure of this effort is the literature produced on the subject. In 1963, the United Nations held a conference on science and technology for development similar to the one proposed. That conference produced 2,000 papers and, altogether, ten million words, summarized in eight volumes comprising one million words. Since then, a dozen small international conferences have produced an additional several million words on the subject. The literature produced by the dozen United Nations specialized agencies and committees, especially UNESCO, and international bodies like OECD, runs into other tens of millions of words. At the country level, while my bibliography prepared in 1966 on "Science and Technology in India" covers 600 items and 3,450 pages, the one on "S&T in the Development of Modern China," in 1974, by Genevieve Dean, lists 311 books and 600 articles. Thus, the number of words on the theme "Science, Technology and Development" is beginning to approximate, in order of magnitude, the number the great physicist Eddington con- sidered, in 1920, necessary in order for a group of monkeys randomly striking on typewriter keys to produce all the books in the British Museum.

Those optimists who wish to conclude from this that the tens of millions of words to be produced by the 1979 Conference may do the trick will be proven wrong. The panel excludes this possibility in its January 1975 proposal by failing to base its approach, objectives, and the agenda on the exploration of such questions as:

− What was the past effort? What were the results?

-   What determined its results or lack of them?

-   How is science and technology changing, and what are the implications of this change for the development and use of science and technology by the undeveloped countries?

-   What are the basic sociopolitical barriers within the less developed countries to the development and use of domestic and foreign science and technology?

-   What is basically right and basically wrong in the practices and policies of the industrialized countries in this area?

-   What can we realistically hope to do in the light of such findings?

Other built-in failure mechanisms can best be identified by considering the proposal in the light of the panel's claim that the 1979 UNCSTD "would be entirely different from the 1963 conference." Such a comparative analysis is necessary in view of the panel's ambition to make the 1979 conference "a turning point in human history," compared to that of 1963 to be "a beginning of vast importance." One crucially important similarity is found in comparing the participation in the two conferences. Just as in 1963, the 1979 effort will be "a conference of governments," represented by the planning ministries, ministerial science advisors, United Nations specialized agency officials, and "eminent experts" and scientists, all of whom would prepare papers and reports for the conference.

The size of the planned conference will be as gigantic as in 1963: at least 2,000 individuals are expected to attend and a corresponding number of papers will perhaps be written. This will make the envisaged effort of directing its preparation and sessions along the rather rigidly envisaged lines extremely difficult, if not illusory.

Even more serious, however, is the fact that, in repeating the participation pattern of 1963, the panel limits severely the exploration of the causes of the past "very limited progress" in the field. For one can hardly expect official representatives of governments and of United Nations specialized agencies "to identify the hindrances and take action to remove them" in relation to the problem arising from the actions of these governments and agencies, except in a very one-sided, superofficial, and biased way. By not providing in its January 1975 proposal any mechanism, nor calling for mutually critical analysis of the past effort, even of the above named two categories of participants – national governments and United Nations agency officials – the uncritical outcome is a foregone conclusion.

Each cell and organ of the world S&T policy community is in a continuous state of more or less sharp confrontation between conflicting interest groups, views, and contradictory approaches. The organized articulation of these views and approaches is the central task in preparing such a conference. I am not suggesting that the participation should be limited to "nongovernments" and United Nations "counter-establishment."   Yet, it seems to me, that the panel, in following the old well-beaten paths, is missing an opportunity to be innovative in the field of world conferences, a need which is widely felt. This could have been done if the panel had proposed specific steps to achieve a joint participation and confrontation of the usual conference representatives with the "counter-conference" ones, and then planned and stimulated the confrontation of their views.

The four basic stated components of the "approach" in the proposal are:

1. "Universal national involvement";
2. "Policy action"; and what I call
3. Global A level; and
4. Global B level.

Global A level calls for "a unified integrated application of the totality of man's potential knowledge (science and technology) to the totality of interacting needs," and Global B level calls for identification of global trends. The last three of these four basic traits of the approach are more or less commonly invoked superficialities, totally divorced from the political realities. The first, "the universal national involvement" is more of the 1963 bureaucratic same. For it calls for "the establishment of ad hoc mechanisms in the participating countries which should oversee the preparation of country papers and promote discussions between officials responsible for economic, sectoral, social, and scientific and technical planning and decision making." The narrowly defined national involvement resulting in "country papers" will form the basis for regional reports "in accordance with guidelines drawn up with the assistance of interregional expert groups," thus ensuring the monopoly of expression of views of a small, selected number of world S&T policy community interest groups, without any provision for the confrontation of views, even among these groups.

Another key built-in failure mechanism in the approach is the nowhere stated, but implied, belief of the panel that a United Nations conference chief objective is to provide solutions to specific problems, and not to organize and carry out interactive confrontations on the identification, definitions, methods of study, and policies for dealing with such problems. This is the impression I gained from the whole proposal and especially from the detailed description of Topic I, where the basic problems to be dealt with are food and agriculture, energy, natural resources, industrialization, and tropical diseases.

In the rest of this chapter, I intend to render even more specific my own views, partially stated in the preceding paragraphs, on the preparation of such a conference. It could be stated that the UNCSTD should result in:

1. broad universal involvement;

2. production of knowledge or intelligence about the problem;

3. interactive critical confrontation among all the participants; and

4. organization of a summary of conflicting viewpoints.

About the first of these, I have given my views in the preceding paragraphs. As regards the second, it is a fact that no international or national organization, or individual, has produced even a first approximation study of either the world S&T indicators of the global R&D potential, or a program related to subjects and social functions, and their regional or country-wide sectoral distribution. Without such intelligence the belief in the "lack of progress" will remain just that. From personal experience, I know that at least two of the most important United Nations specialized agencies do not know, nor do they want to know, how much research and innovative work is being done in their particular field in the world, nor how it is distributed regionally or nationally. Hence, the first task in preparing the conference is to ask every participant, individual, organization, or country to prepare – on the basis of an indicated minimal consensus in respect to the method – the data on basic indicators for S&T development, use, and production, and on relevant policy indicators. The organizers of the conference should prepare and circulate to all the envisaged participants, and to the public, a succinct paper based on their own work on the world S&T potential, S&T program by social functions, S&T production, S&T policy organization, and international interchange disaggregated by regions and countries as related to their degree of development, size, geographic location, and political system. Consequently, the first objective should be to present such a picture and to find ways and means to advance our knowledge in terms of indicators and data of the state of S&T, its changes, and its use.

The second, and equally important, knowledge task is to stimulate the production of intelligence on the changes occurring within the S&T production and social organization, about which crucial subject the proposal has not a word to say. Yet it is only on the basis of preliminary surveys of the present state and trends of change in the development of S&T, and national and international policies and their changes, that one can estimate the presnt day world S&T gap between the developed and less developed countries and begin to discuss the consequences of this gap, as yet totally unexplored.

The interactive confrontation can be stimulated in preparing the conference in three ways: the first is, of course, to ensure a more varied representation of the conflicting views within each country. The second way would be to address to each specific group, or type of participants, specific questions to be dealt with in the papers. Thus, for example, besides a number of questions dealing with the causes of lack of development of S&T and their use within each country, every participant from a less developed country should be asked to answer in his contribution suchquestions as: How much has United Nations special agency X, Y, etc. contributed to the development of S&T capabilities of this effort? How much have national aid agencies contributed on a bilateral basis in the relevant S&T fields in your country, and what are your criticisms of these contributions? In what specific instances and fields do you consider the S&T of the developed countries as counterproductive to the development of your country as a tool of dominance and exploitation? What defenses against it at the national level have been or should be established, and what do you think can and should be done about it at the international level? Another battery of questions should be asked of the developed countries and international agencies to be answered in their contributions. For example: What are, in the light of your experience and studies, the persistent "diseases" of the growth and use of S&T in less developed countries? What is the role of their governments and their science and technology communities in failing to identify, in causing, or in dealing with these dysfunctions? The insistence that similar questions should be asked and answered by the various groups of participants is only the first step to make the conference a forum for interaction confrontation.

The third task of the preparatory committee is to identify and spell out the conflicting approaches to the problem of science, technology, and society, and ask the participants for a critical appraisal of them. Here are only a few from a much longer list of such contrasting approaches. First, S&T imperialism and exploitation versus S&T aid. This subject is being intensely discussed and forms the basis of debate on national S&T policies both in the developed and Third World countries. The issue of the growth of multinational corporations, S&T aid policies and organizations, and the technology-transfer issues have to be brought out from the corridors of the national policy forums, and from public debates, into the light of a United Nations conference.

The second confrontation in the open forum of UNCSTD should take up the descriptions of the present state of interaction among science, technology, and social change prevalent in the two major political parts of the developed world. In the COMECON member countries the science and policy literature, both governmental and nongovernmental, is full of descriptions and analyses of the science and technology revolutions, and the advantages the socialist system offers to the development and social uses of this revolution. In OECD countries, although no single view predominates, the current state of interaction of S&T and social change is spoken of as "the growth of the knowledge industry as related to democracy and multipolarity in its production, use, and control." The confrontation of these two partly contrasting approaches may produce extremely useful insights for the S&T policies of the less developed countries, only if the proponents of the two positions look critically at each other's approaches and experiences before a global forum.

Two strategies of development are now confronted in the world debate: growth through investment and entrepreneurship with open borders versus growth through egalitarian distribution and total social development with selected types of foreign relations with the world S&T. The systematic confrontation of these two views of development can have a very stimulating effect on the national S&T policies of less developed countries.

During the past few years, there has been a fairly active movement stimulating discussion about appropriate, low cost, alternative technology. In the developed countries the movement is oriented toward nonconsumer value systems. Its center, stimulated by writings about the experience of China, and partly of India and some other countries, is mostly oriented toward devising S&T policies which will serve the needs of the majority of the populations in the undeveloped parts of the world. The confrontation of this appropriate technology with the conventional S&T big science and technology and world S&T front policies may be of extreme value in finding the proper mix for various developing countries.

To end this list of possible approaches that have to be openly debated through the proper organization of UNCSTD, I shall mention only "the superpowers" versus "the rest of the world" approaches. The world S&T potential is dominated in the first place by the United States and the Soviet Union, and in the second place by the EEC countries and Japan. The superpower view of the world, of its S&T and relevant policies, is very different from that of a small or undeveloped country, regardless of its political system, but it influences the respective policies of both superpowers and small countries. In this connection, it is worthwhile to mention the need to consider the problem of growing interdependence on a global scale in the field of science and technology, with strong propensity for national science and technology independence.

19 Uncertain Prospects for the UNCSTD— Three Major Underlying Issues

Miguel S. Wionczek

While the general feeling of malaise in respect to the prospects of UNCSTD continues to increase everywhere with the approach of its date, most criticisms are addressed to the institutional mechanisms established for convening the conference. Its Secretariat is widely criticized for "not providing more of the analystical tools required for discussing how to handle science and technology for development.(1) The governments of LDCs are also criticized for not showing much interest in the conference preparations and leaving, in most cases, the work on national papers – expected to represent the core of that global meeting – "to institutions and individuals of little weight in the political, economic, and cultural life of the country."(2) As far as it can be detected, the advanced countries are not too keen on the UNCSTD either, although this aspect of the global situation has as yet escaped critical comments.

Although I subscribe to most of these criticisms, it is not the purpose of this chapter to add more fuel to that particular fire. Instead, I will attempt to address myself to some major, underlying issues responsible for the very uncertain prospects for the UNCSTD.

These underlying issues may be divided into three categories: the present state of the world economy, which largely explains the lack of interest in the UNCSTD on the part of the industrial countries; the sociopolitical background of the LDCs' attitudes towards science and technology, which provides the explanation for the excessive bureaucratization of their preparations for the UNCSTD; and last but not least, the modus operandi of the United Nations system responsible for the way in which the UNCSTD Secretariat functions.

The brief analysis of these three issues strongly suggests that the widely expected failure of the UNCSTD as a forum in which a major advance in scientific and technological relations between the advanced countries and the LDCs could take place has been assured beforehand by the work of forces which cannot be controlled by the UNCSTD institutional mechanisms themselves.

THE PRESENT STATE OF THE WORLD ECONOMY

Let us start with the present world economic situation. While many scientists and some technologists may not see any connection between the world economy, science and technology, and the prospects for the UNCSTD, not only does such connection exist, it is a direct one. It is here where the UNCSTD problems begin.

There is growing evidence that the advanced countries, both in the West and the East, have recently – since 1970 – entered the stage of protracted economic stagnation accompanied by the increasing uncertainty about the rate of future S&T advancement. The present decline in the rate of growth of the advanced countries, referred to by some as an international economic crisis and by others as the beginning of a new era of relatively slow growth; already has affected seriously the growth prospects in most LDCs (excluding a few fortunate oil-producing countries). Even these low rates of growth in the LDCs are being sustained by their increasing financial dependence upon the advanced countries and the oil producing nations. This dependence is reflected in the continuous expansion of the LDC external debt at rates clearly unsustainable in the long run. In brief, in the midst of a major international economic crisis, the dependence of the LDCs on the overall growth of the advanced economies is bigger than ever.

While some radical social scientists suggest at this particular juncture that the LDCs should "delink" themselves from the advanced sector of the world economy and take the road to self-reliance, no one as yet has been able to offer a reasonably coherent political and economic model for the "delinking" and "self-reliance" in the spheres of trade, finance, and technology. The fact that socialist countries – including China – have been expanding their relations with the world economy, because of their needs for more trade, more finance, and more technology, suggests that the design and the implementation of such a model might be extremely difficult.

Certain disquieting developments which feed the uncertainty about the future pace of S&T advance, particularly in respect to industrial innovation, should be considered in the framework not only of general economic stagnation but also of power politics. The attempts of many major Western countries to close the technological gap with the United States, the worlds' biggest economic and technological power, has led to the growing fears in the U.S. of losing its lead in technology to its competitors.(3) Both developments may have most serious negative consequences, not only for technological interchange among the Western advanced countries but – what is more important for the purpose of this paper – for the less developed world.

I wish someone could convince me that the situation is less serious. During my extended stay in Europe in the past few months, I had the opportunity to discuss at length with knowledgeable people of many nationalities the Western S&T problems and policy issues, both at the national and at the global levels. Everybody seems to agree, first, that the present unprecedented level of socioeconomic welfare enjoyed by the advanced industrial societies has been largely – albeit not exclusively– due to the concentration and the exceptional rate of S&T advancement in the West over the past 100 years; and, second, that, as witnessed by the postwar emergence of West Germany and Japan, a causal relationship can be established between the S&T advancement and the international changes in the distribution of political power. But, when the discussion turns to the present and the future, gloom starts to pervade the scene.

There is general consensus in respect to the growing economic and S&T competition among the U.S., West Germany, and Japan. Most of the interviewed people link it with the uncertain future of the world economy and with the internal necessity of each country to defend its share of the world economy which expands presently at a rate much slower than in the past quarter of a century. Many people see in the cards the emergence of technological protection measures, a corollary to the increasing wave of trade protectionism. Some relate, also, the trend towards technological protectionism to the signs of the slowdown in S&T advance which, in turn, seems to be related to the overall economic stagnation. The slowdown in S&T advance should be understood not as the decline in the rate of scientific discoveries but as the slowdown in the application of S&T for the production of goods and services and for the increase in welfare.

As the proofs accumulate to the effect that neither economic growth nor S&T advancement are endowed with automaticity, the preparations for the biggest global encounter on science and technology for development – the UNCSTD – coincide, at least in the industrial West, with deep and increasing uncertainties with respect to the future of growth, the future of science and technology, and, as should have been expected, the future of development as well. What makes it worse, nobody – neither economists nor S&T practitioners – seems to dispose of analytical tools that would explain what has been happening lately both with the world economy and with the global S&T advance.

The implications of these uncertainties for the LDCs cannot be disposed simply with the attitudes, spreading in the advanced societies, of questioning the importance of growth and the relevance of S&T advance. At their present level of income and welfare, the advanced societies may forego economic growth for some time more and live quite comfortably with the slowly expanding stock of S&T, once they learn how to use it in a socially more meaningful way.

Unfortunately, this static recipe does not offer a satisfactory solution for the LDCs' problems. First, the nature of international economic relations is such that the persistent slowdown in the growth rate of the advanced countries will ruin whatever limited prospects the LDCs may have to escape from their present overall social and economic misery. Second, the nature of S&T relations among the advanced countries suggests that under conditions of the secular economic stagnation, and given the dynamics of power politics, the access of the LDCs to S&T may suffer further limitations.

Thus, when some well-informed observers wonder why so little interest is being paid to the UNCSTD in Western Europe and why – as it appears – the United States will go to Vienna without being able to commit itself to any concrete action program for S&T assistance to the LDCs, perhaps the explanation is simpler than it looks.

If the major advanced countries are so concerned with their relative technological positions; if, furthermore, they believe that technology is the key to their economic and, hence, political power; and if, finally, the world economic prospects are highly uncertain, then from the viewpoint of their short-run interests, these countries would have no reason whatsoever to liberalize the access to S&T for the rest of the world and to assist it in the task of building up their own S&T capability. Such negative attitude of the industrial countries would affect particularly the more developed LDCs which have been attempting for some time to increase their share in those parts of the world trade that involve the mastery of modern technology. After all, one might ask: why, in the static international economic situation, would industrial countries be interested in fostering competition when there has already been too much competition? The beggar-your-neighbor attitudes that start raising their ugly heads all over the world would strongly work in favor of not only trade protection but of "technological defense" as well. All this does not bode well either for the LDCs or for the UNCSTD immediate prospects.

LDC ATTITUDES TOWARDS SCIENCE AND TECHNOLOGY

For the purpose of explaining the LDCs' apparent lack of genuine political interest in science and technology on the eve of the UNCSTD, it is necessary to recall how the LDC governments got involved in these matters. The explicit interest of the LDCs, and especially of the more developed LDCs, in science and technology as a potentially important tool for as yet not well-defined development dates only from the early 1960s. It followed in the footsteps of the belated discovery by economists from the advanced countries that technology is an endogenous factor of economic growth and of comparable importance to capital, labor, and physical resources.

The "discovery" of science and technology by the LDCs can be related also to 1) the advance of the industrialization process in the major LDCs, 2) the penchant of the LDC economists for macroplanning exercises, 3) the LDC technocrats' faith in the state's ability and willingness to foster S&T in the absence of interest on the part of the private – both domestic and foreign – sector, and, finally, 4) the influence of some international agencies like UNESCO and UNCTAD (and the OAS in the case of Latin America).

Most probably, each of these factors was in part responsible for the appearance in Latin America and in major Asian LDCs some ten years ago of governmental agencies charged with the design – albeit rarely with the implementation – of S&T policies known by the generic name of National Science and Technology Councils.(4) In the late 1960s, some LDCs bent upon rapid industrialization created additional S&T policy instruments for the purpose of regulating technology imports and modernizing national industrial property legislation. All these initiatives were based on the assumption that the coalition of the state, national S&T communities, and the local private industrial technology users would support policies aimed at expanding domestic S&T capability and relating it to national development needs.

The LDC experiences of the past decade have demonstrated that with very few exceptions (Brazil and India, perhaps) these expectations were too optimistic. The LDC proponents of scientific and technological autonomous — but not autarkic — development found it extremely difficult to mobilize for their cause other major actors, whose counterparts in the advanced countries fully participate in the S&T endeavors: the state, domestic entrepreneurs, and scientific communities. While the state proved to be largely illiterate in respect to S&T problems, and domestic entrepreneurs found nothing wrong in profitable coexistence with technological dependence, weak local scientific communities responded with fears and suspicions to what looked like bureaucratic interference with their liberties and freedoms.

Consequently, National Science and Technology Councils got bogged down into marginal bureaucratic activities; in the absence of political support at the highest level the attempts of countries to engage in S&T planning withered away; the expected rapprochement between the scientific community and the users of S&T did not take place. Not only the private enterprises, but the public sector, as well, continued importing foreign technology indiscriminately and on a large scale, and finally, little happened in respect to the weak and disjointed infrastructure for S&T activities.

The impact of the international economic crisis on the LDCs, which augmented their dependence upon foreign private interests and resources — whether in the form of private direct investment or banking credits — made any progress on the S&T front even more difficult. While in the wake of mounting domestic social and economic difficulties, the state started losing whatever interest it had earlier in the S&T matters, foreign and domestic private interests launched a subtle but vigorous campaign against any public intervention in the S&T domain under the banner of free economy. These developments led to the progressive dismantling of most of the S&T policy instruments in Argentina, Mexico, and Peru over the last few years. They also explain why the majority of LDC citizens who distinguished themselves in the field of S&T policy have abandoned their own countries lately and gone to work for international agencies. In the midst of the general excitement about "brain drain" in the LDCs, nobody has noticed its magnitude in the S&T policy field.

Further evidence supporting these assertions can be found today at any international meeting on science and technology, where most LDC delegates belong clearly to the pre-1960 school of uninitiated diplomats and bureaucrats. This sorry state of affairs has been visible also at the first two UNCSTD preparatory conferences, held respectively in January 1977 in New York and in March 1978 in Geneva. The overwhelming presence there of third secretaries from the permanent United Nations missions is far from accidental. The delivery of the task of writing national S&T papers for the UNCSTD to their local bureaucratic counterparts is another proof of the present political attitudes in most LDCs on science and technology. Facing all sorts of short-run crises and difficulties, the policymakers in the LDCs found such writing expendable. That foreign and domestic private interests helped them in that respect is a matter of record.

UNITED NATIONS SYSTEM RESPONSIBLE
FOR UNCSTD SECRETARIAT

Under these adverse conditions no miracles could be expected from the UNCSTD Secretariat, even if it had all the best experts it needed which are quite scarce in both the advanced countries and the LDCs. Furthermore, the working of the United Nations institutional mechanisms established for the UNCSTD has been handicapped from the beginning by the particular modus operandi of the United Nations system, which — as well-informed people know — has relatively few characteristics of a rational operational system.

It should be kept in mind that the UNCSTD — as all other global conferences — was established on the margin of the United Nations system. So many United Nations departments, bureaus, special agencies, and semi-autonomous organizations deal with science and technology these days that charging one of them solely with this global enterprise would amount to discriminating against all the others, and would upset the delicate balance of bureaucratic power within the United Nations.

In the diplomatic language of the United Nations, all these events can be described as "absence of harmonized United Nations policy on science and technology." One can learn about the degree of this "absence" from a recent report of the ad hoc Working Group on Policy for Science and Technology within the United Nations System, charged in 1976, among other tasks, with assessing whether "any overlapping missions among organizations and bodies within the United Nations constitute healthy pluralism or wasteful redundancies (in science and technology activities)."(5)

According to this report:

1. the United Nations system at present is composed of a wide spectrum of activities in science and technology under widely different systems of planning and implementation and with insufficient coordination among the parts;

2. there is also a need to ensure that greater harmonization, compatibility and complementarity be achieved between policies in the field of science and technology at the national level and those at the level of the United Nations system at large, and

3. in the absence of clear, consensual definition regarding what constitutes a "science and technology" category, no detailed quantitative assessment can now be made of the utilization of the currently available resources for science and technology in the United Nations system.(6)

This _sotto voce_ description of the actual management of science and technology issues within the United Nations system explains, among other things, the intellectually unsatisfactory level of most of the documentation elaborated for the UNCSTD, the disappointing results of the UNCSTD regional preparations meetings, and the degree of confusion surrounding the issue of "science and technology and the future" – item 4 of the conference agenda.(7)

It hardly could be otherwise. Except for the members of the ACAST, many authors of these documents and reports have had very little exposure in real life to actual science and technology _for_ development problems. Furthermore, many bureaucratic interests are in conflict. Consequently, the lessons of the direct experience with science and technology, whether at the level of research or of policymaking, are substituted for by United Nations documents, and "sacred texts" which, by saying little, can satisfy everybody.

SUMMARY

In the absence of "harmonized United Nations policy on science and technology" and in the presence of the attitudes described earlier toward S&T cooperation in the advanced countries, and toward science and technology in the LDCs, it may require inveterate optimism to expect a substantive outcome from the UNCSTD deliberations in Vienna next fall. If things go really wrong, some cynics predict humanity will be compensated, or perhaps punished, by the establishment of a special permanent United Nations agency on science and technology for development.

Does this mean that the reported budget of the UNCSTD Secretariat of $15 million and additional large financial resources that will be spent by the official participants in the 1979 Vienna pilgrimage represent pure waste? Not necessarily.

The preparations for that conference are prompting many groups of concerned citizens all over the world – scholars and nonscholars alike – to put their minds to the questions of science and technology policy in the context of both development and underdevelopment. Given the general state of ignorance still prevalent in this broad and little-explored field, these independent initiatives have already been expanding our scant knowledge about these extremely important global issues. Such initiatives deserve encouragement both before and, particularly, _after_ the UNCSTD, if only because, in the world of power politics and international and national bureaucratic games, free inquiry about the forces that are shaping our future is very badly needed.

NOTES

(1) Jon Sigurdson — Welcoming remarks to the Lund seminar on "Technology, Science and Development in the Changing International System," (May 31-June 2, 1978), as quoted in Lund Letter, No. 6, July 1978.

(2) "National papers — into the final stretch," Lund Letter, No. 5, May 1978.

(3) See as evidence the following excerpt from a long article date-lined Washington, D.C., that appeared in the International Herald Tribune (Paris) on April 18, 1978: "For decades, every new technology or its product seemed to have Made-in-U.S.A. stamped on it, from instant copying and instant photography to advanced computers, nuclear reactors, oral contraceptives, synthetic fibers, and jet airliners . . . Things have changed. There is concern in the White House and Congress, in industry and universities, that the United States is losing its technological lead."

(4) Eduardo Pablo Amadeo, "National Science and Technology Councils in Latin America — Achievements and Failures of the First 10 Years," paper presented to the Symposium (Chapter 11 in this volume).

(5) ACAST, Report of the Ad-Hoc Working Group on Policy for Science and Technology within the United Nations System, E/AC.52/XXIV/-CHP.2, June 23, 1978.

(6) Ibid., pp. 10-11.

(7) See, for example, UNCSTD, Recommendations of Expert Group on Agenda Item 4 of Conference Agenda, EMI/CRP.2/Rev. 2, November 25, 1977.

20 Summary

Vaughan A. Lewis
D. Babatunde Thomas
Miguel S. Wionczek

I would like to make my remarks specifically from the viewpoint of the Caribbean region and the Caribbean researchers on science and technology problems. I start with the assumption that in our discussions on science and technology we are talking about the use of technology for the manipulation of natural environment for essentially two purposes: to provide for the basic welfare of the majority of the population of our countries, and conserve the available resources and to adapt their use to the needs of the future.

The second purpose is particularly important because, in countries subjected to colonialism, it is a tradition to exhaust the known nonrenewable resources in the process of exploitation. This is contrary to the experience of industrialized countries, which have placed a certain emphasis on preservation of their nonrenewable resources and adapted their exploitation to specific ends. In the United States, for example, while the raw materials of all kinds of strategic importance for national security are stockpiled, legislation exists to inhibit the depletion of some resources, like oil, in certain areas of the country.

In the Caribbean, these two objectives, basic welfare and conservation of nonrenewable resources, must be placed in the context of the presence of certain objective factors. The first is the small size of the Caribbean countries; the second, the long history of exploitation of the significant and strategic natural resources in the area; and third, the fact that the Caribbean is composed of open economies, subject to significant mobility of human resources, capital, and natural resources themselves. These factors permit us to define some areas of useful and relevant research.

Let us consider first the question of small size. There has been in the Caribbean for many years a continuing debate about the relevance and significance of small size, whether defined in geographical, population, market, or resource terms. It has been generally assumed in the Caribbean that the small size of the territory, however defined, is an absolute constraint on the capacity for development, of both economic and human resources. In recent times, however, this assumption has been questioned by some economists who argue that size is a manipulative variable, and is particularly so in the context of technological innovation. According to this new line of thought, size represents a significant constraint only when it is looked at in the context of the development and application of particular modes of technology. If this is so, then the fundamental problem for our countries would be the organization of economic and social systems with adequate technologies for communities with relatively small populations – systems that would provide these communities, over time, with basic economic and social security. The viability of such systems would depend on a judicious mix of stability and dynamism, i.e. a capacity for adaptation. Moreover, they also would have to allow for the protection of our communities from unwanted and unwarranted external disturbances.

In brief, in the Caribbean, we need to develop a particular productive system that matches the scale of our countries and their resources. Let us take, as an example, the problem of meeting our basic nutritional requirements. For a long time the West Indies assumed that it was economically viable to export manufactures and to import food. This assumption was based on the belief that an abundance of world agricultural resources precluded any shortage in the world food supply. We have since learned to appreciate that food is both a strategic commodity and a limited resource, and that the continuing need to import it may impose burdens on developing countries. Consequently, it now becomes a basic requirement for Caribbean countries to begin to match their food production system with the size and need of their communities.

On the other hand, our economies have significant natural resources (bauxite, oil, and natural gas, among others), which have been exploited with the use of highly sophisticated technologies largely unknown to our socieites.

Our science and technology policy and, therefore, the research related to, and deriving from, that policy, must operate from two perspectives simultaneously: 1) we must learn how to solve the technological problems of providing basic necessities to our people, and 2) at the same time, create policies to deal with sophisticated technologies used by outsiders who exploit our natural resources. We have to deal simultaneously with the polar extremes of a broad technological spectrum. This puts a tremendous strian on our supply of available human resources which is limited, partly, because of the size of the population. However, the Caribbean researchers need to understand the behavioral dynamics of foreign firms exploiting our resources for export abroad, as well as the dynamics of their technological progress and experimentation. For a small country, or a group of small countries, this is a major task with which we must come to terms. In addition, our science and technology policies must somehow reconcile the acquisition of technology needed to exploit our resources with the limited funds our small economies can supply, in the context of rising prices and expanding international trade. We need to recognize and understand not only the technology, but also the political dynamics which contribute to technological innovation.

This takes me into what I mentioned as a third objective factor: the mobility of the factors of production. Historically, the movement of human resources and capital has been a significant variable in the development of economic and social systems. The implication of this is that we need to develop a capacity to hold onto our human resources capability, once developed and trained. The brain-drain does not exist in a vacuum. Certain types of technological and experimental activities have traditionally taken place in some areas and not in others, and, as a consequence, there is a tendency for trained personnel to drift from one part of the world to another. This drift is determined by the location of the institutions that work on technological innovations and the particular framework of relations in what those institutions operate.

To summarize, I would like to say that, in the Caribbean, we should concentrate our research work on the provision of basic needs, on natural resources, and on a science and technology policy related to the small scale of our economies.

– Vaughan A. Lewis

* * *

My comments are going to be very general and brief. To any participant in our symposium it must be apparent that we have moved towards a shared and mutual perception of the problems and issues relating to the building of scientific and technological capabilities in LDCs. This mutual perception is still constrained by some residual difficulties of definitions with respect to the interrelations between science and technology, and the joint use of these two important activities for the purpose of development.

Since a great amount of empirical data has been presented in the papers written for this symposium, it should be incumbent upon us to use these specific analytical attempts to explore the various viewpoints on critical issues. A first step is to see what can be done about arriving at a common terminology in order to clarify our thinking and obtain more information. Listening to the presentations, it sometimes seems that each country, and each region, faces unique problems. However, the symposium offered evidence that the basic problems of fostering scientific and technological efforts in LDCs are similar everywhere.

The fundamental question, raised several times during the seminar, was put by one of the participants, Mario Kamenetzky, in the following terms: "In trying to address all the problems related to science and technology for development, one critical question that we really need to address is: What kind of society do we want?" I think that this is the starting point for research. Where do we go or where do we want to go? What are the vehicles? Which are the viable ways of developing a reasonably just society? And what are the resources we need to achieve such a goal? How many of these resources do we have or could gain access to reasonably, and what is their quality? These are very important questions to which we need to address ourselves.

One our mutual concern is expressed, it becomes necessary to be policy-oriented and practical. We need still more data on S&T performance, and on the obstacles to S&T progress in the LDCs. This point has been emphasized over and over throughout the symposium. We need data from in-depth country studies, from which we can derive specific conclusions with respect to many factors, such as the degree of political support, educational level, economic development patterns, and cultural and regional diversities, all of which affect the level, the scope, and the quality of scientific and technological activities in the LDCs.

Florida International University is in a unique position to contribute to the necessary efforts in these fields because of its geographical location, its organizational structure, and its interest in science and technology problems. One critical area into which we might move, here at Florida International University, is to conduct research on the feasibility of effectively utilizing nonproprietary technologies in LDCs. With the feasibility ascertained, attempts could also be made to identify the technologies which are applicable in specific LDCs and can be channelled to small-scale businesses in both the agricultural and the industrial sectors. Major research in this area is critical to the success of any practical program designed to deal with the immediate problems of food and shelter in most LDCs. It is, therefore, essential that this particular subject receive major attention in the preparatory deliberations of UNCSTD. There is not doubt in my mind that, in order to alleviate some of the pressing problems of low productivity and establish a viable foundation for future development in the LDCs, attention in these countries must be focused on the hard realities posed by the availability and the rate of development of physical and human resources. The policies and practical methods of dealing with these realities must necessarily include knowledge of access to, and application of, technologies in LDCs, (proprietary and, in particular, nonproprietary technologies) for the production of producers' goods, and for the production of food and shelter – both of which are critical to the lives of two-thirds of the world's population.

– D. Babatunde Thomas

* * *

Contrary to some schools of thought, I believe that the LDCs need not only technology, but also science, not only because of the direct relationship between scientific and technological activities, but because science plays an important social function which can not be forgotten for the sake of pragmatic short-run considerations.

Furthermore, I believe that, for historical and other reasons, the S&T problems in the context of underdevelopment are very much different from those faced in the advanced countries, whether capitalist or socialist. When people in the LDCs talk about the "new science" or "new technology" for the LDCs, they have in mind two factors that make it advisable to orient the S&T efforts in the LDCs in a different manner from that followed in advanced societies. These factors are: distinct social and economic structures, and differing S&T needs.

There is still a lot of misunderstanding in the advanced countries about the developmental role of science and technology. The problem that faces the LDCs in that respect is not only how to strengthen and expand research and development activities (R&D) but how to reorient them towards real and urgent social needs. Such orientation seems impossible in the absence of an infrastructure for science and technology, and of permanent links between the S&T system (which by definition includes the infrastructure) and the educational system, on the one hand, and the productive system, on the other. These problems do not appear in the developed societies where, over the last two hundred years following the first industrial revolution, both the infrastructure and the links have been established gradually. Their absence or weakness in most, if not all, of the underdeveloped world makes science and technology planning in the LDCs a necessary prerequisite to meaningful scientific and technological advancement.

Moreover, the LDCs face a situation in which the knowledge and technical know-how supplied to them by advanced societies is very often unsuitable for the solution of problems of underdevelopment. This is so not only because of structural differences between the advanced and the underdeveloped world, but also because of the dynamics of S&T advancement in today's postindustrial societies. Not only are the LDCs very far from that development stage, but they have not as yet become industrial societies. LDCs can be defined as preindustrial societies, endowed with a very thin top layer of "modernization" and "industriali-zation." Consequently, they need not only industrial technologies, but also the knowledge and the know-how suitable for the solution of nonindustrial problems. It would be highly advisable if people dealing with science and technology policy issues in the advanced world would keep this in mind when they attempt to design S&T solutions for the LDCs. Recently, in a very interesting paper on the subject of appropriate technologies, the Canadian economist, Rybczynski, cor-rectly pointed out that most of the work about appropriate, or intermediate, technologies, along the lines of "small is beautiful," has originated in the advanced world. This infatuation of the DCs with "small is beautiful" can be clearly linked to the disillusionment of many people in advanced societies with the working of industrial societies. While such disillusionment can be easily understood, most LDCs have still not had a chance to become disillusioned with the consequences of industrialization, if only because they have not yet gotten to that point.

Speaking about the research needs of the LDCs, one should make a distinction between research on substantive S&T issues, and research on science and technology policy. Of course, in Latin America and other underdeveloped regions, we need both; but what should come out of this meeting is a set of suggestions with respect to what could and should be done in the science and technology policy field. I think that the list of things to be done on both levels is very long.

One may take, as a point of departure, the list of obstacles to scientific and technological advancement elaborated in March 1977 at the second preparatory meeting for the United Nations Conference on Science and Technology for Development (UNCSTD). The list, arrived at by consensus between DCs and LDCs, identified three types: national obstacles, regional obstacles, and international obstacles. In UNCSTD the emphasis is mainly on international obstacles. But, as we have heard often in our seminar, a great number of obstacles exist also at the national level. These are no less important than international obstacles to the advancement of science and technology. The major internal obstacle should become the subject of multidisciplinary research projects, which might be taken up jointly by people from both the advanced and developing countries.

One topic of great relevance is how (under what conditions and to what extent) the multinational corporations could become a vehicle for real and useful technology transfer to the LDCs. Progress will depend, however, on whether we will be able to abandon the conventional wisdom approach. We need to study specific aspects of technological behavior of MNCs in the less developed countries, in order to discover why the multinationals do not engage in R&D in developing countries. Is such a situation absolutely unavoidable? Can the MNCs be persuaded to do in the LDCs R&D suitable to the needs of the latter?

People from the DCs must also undertake critical studies of the results of international scientific and technological cooperation, looking, in considerable detail, into those domestic obstacles in the DCs that reduce the value of such cooperation to the LDCs. Some work is being done in that respect presently, under the auspices of Pugwash Movement for Science and World Affairs.

When analyzing the problems of LDCs, both in substantive terms and on the S&T policy level, more participative research is needed because most of the research on the LDCs done by the DCs alone is just not relevant. It is not because the DC researchers are incompetent, but because they do not sufficiently know the societies which they attempt to study, and to which they then attempt to give advice on policy. Obviously, it is not enough to stay for six months at the Cairo Hilton and merely have contact with other foreign visitors, and the embassies, to understand the working of the Egyptian society. Neither will collecting all possible quantitative data and putting them through computers do the trick. Moreover, understanding the problems of the underdeveloped societies is not an easy task for people who have been trained, and sometimes overtrained, in only a single discipline; they often have great difficulty in facing problems involving disciplines other than their own.

What is also absolutely necessary for scientists and technologists in the developed countries to realize is that scientific communities do exist in many LDCs. These may be small, and not particularly well integrated, but they may see the science and technology requirements of development problems in a somewhat different way than their colleagues in the DCs.

Finally, there is a considerable stock of knowledge in the LDCs about such subjects as international technology transfer and its present shortcomings. Unfortunately, most of that information is unknown in the developed countries because of their highly ethnocentric attitudes. Unless such attitudes in the DCs are abandoned, it is difficult to expect much progress in the field of international cooperation in science and technology matters, and, hence, in S&T advancement in the LDCs.

— Miguel S. Wionczek

A REPORT ON THE SYMPOSIUM

D. Babatunde Thomas

INTRODUCTION

This report is a summary and an analysis of the deliberations of the April 6-8, 1978 Miami Symposium, which culminated in this volume. The key issues, the conclusions, proposals, and follow-up activities in the form of research projects, are outlined. The deliberations of the symposium, as published in this volume, have been partially supplemented by three additional papers.

The symposium was convened for nongovernmental participants to examine various facets of the problems of scientific and technological development in LDCs, with a regional focus on the Caribbean and Latin America, and to address the preparatory activities associated with the United Nations Conference on Science and Technology for Development (UNCSTD), scheduled for August 1979 in Vienna, Austria.

Based on this agenda, the symposium focused on what the critical scientific and technological issues are in solving the development problems in LDCs, and raised questions about the possible role of UNCSTD in charting viable approaches to help solve these problems through appropriate use of science and technology.

THE KEY ISSUES

The key question addressed by the symposium was the following: How can science and technology be more effectively employed to deal with the problems of basic human needs, such as food and shelter, and to accelerate development?

The main issues in recent discussions on the development of science and technology in LDCs were typically couched in terms of these countries' needs. Namely,

1. What needs and resources do LDCs have in order to build and organize their national scientific and technological capabilities?

2. What forms of assistance are useful in meeting these needs, and how can these countries obtain the requisite assistance from

each other and from advanced countries on mutually agreeable terms, subject to changing circumstances and the structural requirements of the individual countries?

3. How can advanced countries assist in increasing the flow of technology to meet the needs of LDCs?

4. How can advanced countries assist in increasing the flow of appropriate technologies to meet the needs of LDCs?

Clearly, many technological needs in LDCs can be met by the importation of technology, in a much shorter time frame than by indigenous development. However, imported technology and its specific forms are usually not appropriate unless thoroughly adapted to local conditions. Appropriate technology need not be rudimentary or tradi- tional. Appropriateness, therefore, must be defined in terms of the needs and the circumstances of the user country, industry, or firm. Production technologies should not only meet the needs of the productive sector, but should also be adapted or designed to make optimal use of available local inputs and raw materials. Technologies should have a sound scientific basis, be inexpensive, effective, and capable of producing competitive products and services. Appropriate- ness of technology is a subject of universal application. It has been suggested that most of the work being done in recent years on "appropriate technology" is a consequence of the disillusionment with the environmental impact on the workings of industrial societies rather than out of concern for the problems in LDCs.

For emphasis, the foregoing issues and questions can also be examined alternatively from a more specific viewpoint; namely,

How can LDCs improve the terms under which technology transactions are negotiated with private and public technology suppliers from advanced countries?

Before answering the first of the questions outlined above, the leadership in individual LDCs must first answer a more basic question; namely, what type of society does their nation want to build? The answer to this question can be expected to reflect a social opinion influenced in part by the social values which the political leadership promotes.

Rather than devote its entire time to fashionable topics, such as technology transfer, appropriate technology, and the like, the sympo- sium concentrated primarily on policy aspects, planning, and other strategies for the creation of national and/or regional scientific and technological capabilities in LDCs. This is not to suggest that the conceptual and operational problems of technology transfer and appropriate technology are not germane to the building of indigenous scientific and technological capabilities. By the same token, the transfer of technology, in and of itself, is meaningless if it is not discussed, analyzed, and implemented within a context of establishing the necessary internal capabilities.

CONCLUSIONS AND PROPOSALS

In the search for concrete proposals, the symposium focused on four areas:

1. viable approaches to building scientific and technological capabilities;

2. infrastructural support;.

3. technology transfer/acquisition; and

4. science and technology policy and planning based on national circumstances.

The symposium focused on these issues, both with and without consideration of how they are being addressed in the intergovernmental preparatory meetings of UNCSTD. Unlike the latter meetings held thus far, symposium participants did not attend as national representatives, but as individual scholars and professionals. Consequently, exchanges and contributions made in the sessions were not constrianed by protocol; but, rather, were frank and blunt in their analyses and criticisms of the prevailing international order in its economic, social, and political dimensions.

From the four areas outlined above, the symposium concentrated its deliberations on the following six points.

I. Building indigenous scientific and technological capabilities in LDCs is a long-term proposition. The successful building of these capabilities requires infrastructural planning and development which entails 1) a major commitment to reorganize and upgrade both the level of education generally and that of science and technical education in particular, as well as to improve the quality of research facilities; 2) developing the capability to manage scientific and technological research; 3) a major improvement in the extent of political patronage; and 4) monitoring the allocation of resources for the development of science and technology to insure the social utility and operational efficiency of the resultant projects.

In building a climate supportive of science and technology, for example, by improving the general level of education, particularly science education, the application of scientific methods of society's problems must be encouraged at every stage of educational training and development.

To encourage political patronage, scientific and technological research activities must have high visibility, for example, focusing on top priority problems in the satisfaction of the needs of the majority. High visibility attracts support and encourages the political leadership to broaden its patronage of science and technology projects, and promote policies to institutionalize science, not as a luxury but as an integral part of the requirements for national development. An important fact to be noted is that major decisions concerning development of scientific and technological infrastructure are made, or significantly influenced, by political leaders. It should be noted also that the decisions entailed are much too large to be made by any single group.

Mechanisms should be developed to enable scientists, engineers, technologists, technicians, and entrepreneurs to interact and hold key roles that involve decision making for the building of national S&T capabilities.

II. There is a critical need for empirical investigations into the role of multinational corporations (MNCs) in the international transfer of technology, and the contributions they make to the building of indigenous capabilities in LDCs. The necessity for such an investigation is based on the premise that many of these corporations are capable of supplying LDCs with essential input requirements in the building of technological and managerial capabilities.

MNCs have patents and blueprints but, more importantly, also have production technologies which have been tested and perfected through practice. However, experience in LDCs suggests that most MNCs supplying technology do not intend to facilitate effective acquisition, and leave little room for free choice by the purchaser. Among other conclusions drawn from experience are 1) technology transfer by MNCs to LDCs is not an automatic or obvious activity simply because of the latter's physical presence. What seems obvious is that MNCs transfer technologies which benefit the elites, utilizing available local technological capabilities, and concentrate their activities exclusively in the modern sector. 2) The conclusion that multinational corporations generally transfer a minimum of technology to LDCs, a thesis supported by heuristic arguments and some empirical evidence, has validity; but further research must be conducted to evaluate available empirical data, and ascertain the extent to which this conclusion is sustained by the data. The determination on MNCs to treat the technologies in their control as private property supports this conclusion, and imposes major limitations on the supply and acquisition of technology. Furthermore, this determination on their part is a key issue in the debate on incentives and constraints on the effective international flow of technology. The following questions arise: Is the quantity, quality, and the total impact of "transferred" technology comparable to the total price paid, and the incentives provided by recipient LDCs? What are the social and private costs/benefits of the transferred technology, from the standpoint of the demanders and suppliers, and how are any differences to be reconciled?

In order to determine and evaluate the answers to these questions, there must be clear definitions of the needs of the LDCs, and a carefully established and well-articulated set of priorities. The ability to make the specific technological needs of LDCs known to potential suppliers requires the existence of a sufficient number of active national personnel with the managerial and organizational capability to use the scientific and technological capabilities being developed.

III. There is a need to promote the use of nonproprietary technology and develop approaches to improve access to its use by small and medium scale producers. One of the foci of this objective should be the proper design and implementation of programs for effective international transfer, acquisition, and utilization of nonproprietary technology in rural areas and in lagging sectors of the economy.

In formulating policy in LDCs, priority should be given to the establishment of clearly articulated and viable policies to harness nonproprietary technology for development.

Knowledgeable manpower is also required in searching for, and securing, nonproprietary technology.

IV. The significance and relative merits of the holistic and the systems approaches to development and implementation of science and technology policies in LDCs needs to be well understood by policy-makers, particularly in Latin America, given its experiences.

The holistic approach emphasizes the need to coordinate the goals and implementation of science policies with technology policies, so that each works to complement and reinforce the other, rather than each operating in its seperate rarefied atmosphere. However, too often in Latin America, the centralization of policy and decision making apparatus fails to coordinate these functions, but treats them as interchangeable parts, thus undermining the effectiveness of both policies. What is needed is a sensitivity to the unique characteristics of each, as well as an awareness of the ways in which they relate to each other.

The complete centralization of decision making in science and technology development is proving unworkable, in the same manner that centralized and comprehensive planning of the national economies have proven unworkable. National economic plans in LDCs, many of which have been and are being formulated by economists from advanced countries, often fail to consider the significance of the social and cultural environment in planning for national development, and take for granted the requirements needed to develop indigenous scientific and technological capabilities. These factors have contributed to irrelevant responses to the needs of LDCs, and to failure to attain the goals and objectives set by these plans.

There is a critical need to devise new methods of integrating scientific and technological development goals, as well as new instruments by which these policies can be implemented. In the attempt to balance scientific and technological development on the one hand with the requirement of science and technology and national development on the other, it is essential to recognize the nonlinear relationship between science and technology. The links between science and technology are not as smooth and continuous as commonly believed. Furthermore, technological development is not always a balanced process.

V. There is urgent need to integrate scientific, technological, and management decisions in LDCs through the establishment of links among the science and technology programs, the institutions responsible for the development of S&T, and the domestic production sector. The specific links should be two-fold 1) between institutions of higher learning (especially the faculties of science, engineering, technology, and management) and the producers of goods, and 2) between the commercial suppliers and users of scientific knowledge, technological, and managerial/organizational know-how. Consulting and engineering services make a vital contribution to these links and aid in the building of the desired scientific and technological capabilities. Often, the largest users of local consulting and engineering services in the LDCs (for example, in Mexico and Argentina) are the subsidiaries of MNCs. These services should be structured to stimulate demand for science and technology by local businesses in the productive sector. Consulting and engineering services can also act as intermediaries between the foreign supplier and the local user of imported technology.

In planning for science and technology, it is extremely important to be aware of the time limits of the people in political office, and to solicit support accordingly. This becomes a serious matter if the terms of office of supportive political leadership terminate too soon, and thus serve to undermine the long-term requirements of building scientific and technological capabilities.

Experience in Latin America suggests the need for caution in designing an S&T policy so as to avoid conflict with the other problems of underdevelopment. The establishment of CONACYTS in Latin America served to institutionalize science and technology, but there is no conclusive evidence that their existence responded to explicit social demands. The functions of these organizations include the formulation of science and technology policy, planning, and coordination of scientific and technological activities. Their functions are usually directed so as to suggest the natural integration of scientific and technological activities.

This experience indicates the need to establish technical information services designed to assist the productive sector (rural and modern) to identify, define, and solve problems; to enhance profit opportunities; stimulate improvement and innovation; identify and select appropriate sources of technological information; analyze, evaluate, and repackage information in a way understandable to small business firms, and potential entrepreneurs; assist with the utilization of information; promote the use of existing technological resources, for example, laboratories and consulting and engineering services; motivate business firms to seek out ways and means of improving and adapting technology; influence a favorable environment for technology transfer and innovation; promote the use of information in strategic planning; and assist national firms in gaining access to the external technological environment.

Surveys and, where feasible, inventories of scientists, technicians, engineers, and related managerial manpower should be conducted and up-dated regularly. Many individuals with this training experience are currently either unproductively engaged or are underemployed in national and international bureaucracies. Programs should be developed to identify them and incentives designed to attract them back into laboratories and into the productive sector. These would require setting priorities in order to provide them with adequate resources, a free creative environment, and, through national political patronage, encourage, motivate, and orient them towards priority projects that meet specific national needs.

Scientific research groups with similar interests, which transcend their institutional affiliations, should be established, and encouraged to help lay the foundation necessary for building a domestic scientific capability. This requires placing the role of science in a proper and valued perspective, and the institution of appropriate links with the productive systems. This perspective should emphasize the necessity for conducting basic research, in view of its importance in building specific research capability, which can be applied to indigenous problems.

It was recognized that the development of skills for the use of existing knowledge carries a higher priority than the creation of new knowledge; nonetheless, the usual assertion that LDCs do not need to conduct basic research, but need only concentrate on technology development, is antithetical to the building of indigenous scientific and technological capabilities. It is through the development of basic research that some of the specific problems of instrumentation and direct responses to basic needs can be met. Such research must meet international standards in order to promote close interactions with the international scientific community, and also to elicit international recognition and support in the form of funding for other research projects.

In planning science and technology development, distinctions should be made between flexible guidelines and firm policies. A major problem with science and technology development in many of the countries in the Circum Caribbean/Latin American region, as well as in many other LDCs, is the preoccupation with science and technology as ends in themselves rather than as means to an end. Science and technology have to be viewed as tools of development.

Conclusions Relating to UNCSTD

The foregoing points and the policies recommended to deal with them are relevant, urgent, and desirous of top priority consideration by UNCSTD. Although the issues outlined have been duly recognized by UNCSTD, the success of the conference can be measured only by its response to the requirements in the LDCs for their scientific, technological, and economic development.

In this respect, the needs of the LDCs must be unambiguously defined and emphasized. Priorities should be attached to these needs as perceived by the majority population, not just as perceived by the elite. Furthermore, these needs and priorities must be articulated by the political leadership such that they are well understood by the ruling elites in the advanced countries and by the majority of their population. Historically, these needs and priorities, as understood by advanced countries, have rarely, if at all, been the same as the LDCs'.

Incentives to, and constriants upon, the international flows of technology must be clearly outlined from the viewpoint of both the suppliers and users of technology, and attempts should be made to reconcile differences as much as practicable. In the absence of conclusive empirical data, preliminary attempts should be made in this outline to identify and discuss regional, national, and international obstacles to the flows of technology, in particular, those that pertain to cross-cultural and technical exchanges among small farmers, small entrepreneurs, cottage manufacturers, and small businesses. A program of actions to minimize the constraints, and to improve the incentives, should also be outlined.

The conference must come up with concrete proposals and programs, and with commitments to clearly establish time frames: to develop infrastructures in LDCs, to improve productive capacities, and to alter approaches and terms of technology transfer in accordance with the clearly expressed needs of most of the LDCs.

One of the more unfavorable aspects of technology transactions between advanced countries and the LDCs, from the standpoint of the latter, is the supplier-initiated technology "package" in which the recipient is compelled to accept components inapplicable to its needs in order to gain access to the desired technology. "Unbundling" of technology would allow for greater freedom of choice and, presumably, reduce cost. Most importantly, this approach ought to promote versatility in the use of imported technology, and allow choices to be made on the basis of compatibility with existing systems, as well as facilitate adaptation to local needs. But unbundling requires the existence of local technological capabilities for selecting and evaluating technologies, and for the performance of preinvestment activities.

Responsibilities and commitments must be clearly defined, and the implementation of projects and programs resulting from all proposals must involve the full participation of the principals.

Among the goals of LDCs is that of reducing their dependence on foreign technologies. However, this presents a number of problems. There is some confusion between developing indigenous technology and indigenizing imported technology. These are two entirely different issues. The latter is a short-term problem, and the former is a long-term proposition. In both time-frames, the requirements are for a dynamic production technology with problem-solving capabilities, and the capacity to stimulate innovative activities, as well as to conserve scarce resources. Different approaches to the problem of reducing LDC dependency upon technology from the advanced countries were discussed, but no consensus emerged. One approach suggested that reducing such dependency requires the cooperation of the suppliers of technology, primarily MNCs, so that LDCs are able to negotiate for a fair price, and to secure guarantee of unrestricted acquisition and utilization of imported technology. In addition to the transfer of technologies and obtaining assistance with their redesign and adaptation, it was also recommended that avenues be explored to involve MNCs in the development of new technologies applicable to the expressed needs of the LDCs. Technologies most susceptible to reducing LDC dependence are those that can be adapted to local conditions, and that facilitate the development of local technology. Still another approach feels it is unnecessary for the MNCs to become involved in local scientific and technological research. Rather, it suggests that the technological needs of the rural and poor sectors be met through indigenous efforts, and appropriate technologies be acquired through foreign sources (for example, through MNCs) only for the development of the modern sector. Finally, it was recommended that adaptation and use of modern technology should take account not only of short-term needs alone, but also of the long-range requirements of development.

Research Projects

The following is a sample of research projects for follow-up activities by the symposium.

1. Collect data and conduct an empirical study on the processes of technology, transfer, and investigate case studies (specifically of rural and small – scale entrepreneurs) to determine requirements for successful transfers. Develop illustrative cases to ascertain constraints on effective transfers, propose guidelines and policy recommendations to eliminate constraints and promote incentives for successful transfers.

2. Conduct an empirical study of the viable alternatives to the linking of educational, research, and productive systems for indigenous development and management of scientific and technological infrastructure in LDCs.

3. Investigate why multinationals do not engage in major research and development activities in LDCs. Determine the basis used by MNCs in setting their research and development priorities, and design a program of action, from case studies, to increase MNCs' conduct of R&D activities in LDCs, and to investigate their costs and benefits for policy recommendations.

4. Study the scope of, and design viable mechanisms for, technological cooperation among LDCs.

5. Ascertain the extent of laboratory research in LDCs (for example, in the area of testing and control), and estimate its direct and indirect effects on development of the scientific and technological infrastructure.

6. Study the role and advisability of using tax incentives designed by the advanced countries and the LDCs to encourage effective flow of know-how and skills from small-scale businesses in the advanced countries to their "backward" counterparts in the LDCs.

7. Investigate the role of financial institutions, in conjunction with consulting and engineering services, in establishing appropriate linkages between supply and demand for scientific and technological inputs for development.

8. Study technological problems related to stages of development and sizes of countries.

9. Conduct a multicountry study of technology planning for basic needs, with a focus on identifying viable means of developing structural shifts in production patterns towards meeting the needs of the general population.

The pursuance of the follow-up projects is in keeping with Florida International University's mission of responding to the needs of LDCs through research activities.

Initiation of work on these projects is expected to serve as the basis for the institutionalization of these needs, and for continuity of positive responses and commitments to them. The institutionalization is envisaged as a center or institute with a primary focus on:

1. development of appropriate methodologies for empirical research on the effectiveness of proprietary and nonproprietary technology flows from advanced countries to LDCs;

2. building scientific, technological, managerial, and organizational capabilities in LDCs, especially in traditional and lagging sectors;

3. improving LDCs' knowledge of those factors which dictate the character and scope of international technology flows;

4. improving advanced countries' knowledge of LDCs' science and technology needs;

5. exploring opportunities for the supply of technology to LDCs by small and medium scale industries, businesses, and farms in advanced countries;

6. providing guidance to policymakers on the formulation and implementation of science and technology policies in small nations with limited markets.

Approaches to these projects will include seminars, conferences, workshops and symposia; collaborative and joint projects with scholars/researchers in LDCs; instructional programs involving short-courses tailored to the needs of policymakers and technicians in LDCs, as well as for policymakers and business executives from the United States and other DCs; publication of periodicals, monographs, and functional technical journals to diffuse information about programs and the results of research projects.

Discussions have begun on questions about Florida International University's internal capacity to institutionalize the suggested activities and to conduct the related projects, and about the scope of the projects, the extent of external support to be expected, and the nature of possible linkages with institutions having a similar orientation, both elsewhere in the United States and abroad.

Index

About the Editors
and Contributors

D. BABATUNDE THOMAS is Chairman, Department of Economics, and Associate Professor of Economics and Technology, Florida International University, Miami, Florida, and formerly a Research Fellow at the Nigerian Institute of Social and Economic Research, University of Ibadan. Dr. Thomas is a frequent contributor to international conferences on the subject of science, technology, and economic development, a consultant, a contributing editor, and the author of two books and several monographs and articles.

MIGUEL S. WIONCZEK is Senior Research Associate at El Colegio de Mexico, Mexico City, former Deputy Director General of Mexican Science and Technology Council, Chairman of the UNCTAD Commission on Technology Transfer, and author of many books on economic development, regional integration, and science and technology policy.

R.O.B. WIJESEKERA, Special Consultant, National Science Research Council of Guyana.

RODRIGO ZELEDON, President, National Research Council for Science and Technology, and Professor, University of Costa Rica.

MARIO KAMENETZKY, Consultant, Office of Science and Technology, World Bank, Washington, D.C.

MIRA WILKINS, Professor, Department of Economics, Florida International University, Miami, Florida.

TREVOR M.A. FARRELL, Department of Economics, University of the West Indies, St. Augustine, Trinidad W.I.

J. DELLIMORE, Department of Chemical Engineering, University of the West Indies, Cave Hill, Barbados W.I.

MAURICE A. ODLE, Director, Institute of Development Studies, University of Guyana, Georgetown, Guyana.

STEVE de CASTRO, Department of Economics, University of the West Indies, Mona, Jamaica.

JAIME LAVADOS, Faculty of Medicine, University of Chile, and Visiting Professor, Florida International University.

EDUARDO AMADEO, Consultant, formerly with the National Council on Science and Technology, Argentina.

JOSEPH B. HODARA, Senior Research Associate, Interdisciplinary Center for Technological Analysis and Forecasting, Tel Aviv University, Israel.

JORGE ARIAS, Head, Scientific Development Division, Instituto Centroamericano de Investigacion y Technologia Industrial.

FRANCISCO AGUIRRE, Deputy Director, Central American Research Institute for Industry, Guatemala.

ARNOLDO K. VENTURA, Chairman/Director, Scientific Research Council of Jamaica, Kingston, Jamaica.

MICHAEL J. MORAVCSIK, Department of Theoretical Sciences, University of Oregon, Eugene, Oregon.

STEVAN DEDIJER, Research Policy Program, University of Lund, Lund, Sweden.

VAUGHAN A. LEWIS, Director, Institute of Social and Economic Research, University of the West Indies, Mona, Jamaica.

Pergamon Policy Studies